Astronomy with your personal computer

ASTRONOMY WITH YOUR PERSONAL COMPUTER

PETER DUFFETT-SMITH
University lecturer in Physics,
Mullard Radio Astronomy Observatory, Cavendish Laboratory,
and Fellow of Downing College, Cambridge

SECOND EDITION

CAMBRIDGE
UNIVERSITY PRESS

PUBLISHED BY THE PRESS SYNDICATE OF THE UNIVERSITY OF CAMBRIDGE
The Pitt Building, Trumpington Street, Cambridge CB2 1RP, United Kingdom

CAMBRIDGE UNIVERSITY PRESS
The Edinburgh Building, Cambridge CB2 2RU, United Kingdom
40 West 20th Street, New York, NY 10011–4211, USA
10 Stamford Road, Oakleigh, Melbourne 3166, Australia

First published 1985
Reprinted 1986, 1987
Second edition 1990
Reprinted 1997

Set in Times

A catalogue record for this book is available from the British Library

Library of Congress Cataloguing in Publication data
Duffett-Smith, Peter.
Astronomy with your personal computer / Peter Duffett-Smith –
2nd ed.
p.   cm.
Includes bibliographical references.
ISBN 0 521 38093 6. hardback
ISBN 0 521 38995 X paperback
1. Astronomy – Data processing. 2. Microcomputers. I. Title.
QB51.3.E43D83   1990
520'.285'.416 – dc20   89-17444   CIP

ISBN 0 521 38995 X paperback

Transferred to digital printing 2003

to Jennifer and Mary and their porridge

CONTENTS

Contents

PREFACE

The microchip revolution has made mathematical positional astronomy easy. Computing power, such as would have been the envy of even the most prestigious institutions several decades ago, is now commonplace in the home. Amateur astronomers can perform complex calculations with comparative ease, making predictions of forthcoming celestial events or demonstrating what took place in the past, with a precision that is almost unlimited.

Many books have already been published which list the steps (more or less clearly) to be followed to the desired answer. Some are for calculators, but others exploit the special powers of the microcomputer. Unfortunately, those in the former category do not provide programs which can be used immediately on the computer, while those in the latter category are often too specifically biased towards particular makes of machine and particular calculations. While it is always possible to modify a program to work on another model or to make a different calculation, it is not always easy to do so, especially when detailed notes about its workings are not supplied.

This book is designed for the person who wishes to use his or her personal computer to make astronomical calculations with the minimum of fuss. It is not specific to any make of machine, neither are you confined to specific calculations. Rather, you are presented with a set of linking subroutines, each carrying out a specific task, which you can combine to form a complex program designed to your own requirements. You need only have a broad understanding of the problem; the subroutines themselves take care of the details.

I have tested the subroutines with many examples, comparing my answers against *The Astronomical Almanac* and other publications. Although I believe them to be relatively free from errors, I am quite sure that 'bugs' will come to light through extensive use, and I would be most grateful to hear of any which you find. I can then correct future printings of the book.

In this second edition, I have included several new routines and more handling programs in response to requests from readers of the first edition. I have also made extensive modifications to all the routines, so that the new ones

Preface

are incompatible with the old ones. This was unavoidable, if only because I have had to limit the length of the name of every variable to one or two characters. If you have used the first edition and now wish to update your programs to take account of the new features of this edition, I'm afraid that you will need to enter every routine again from scratch. You may not wish to bother. However, these new routines have benefited from the comments supplied by readers of the first edition, and are therefore much improved. For example, the routine TIME (which replaces the old routine GTIME) now takes account of the observer's geographical longitude, time zone, and daylight saving correction, to make conversions directly between local civil time and local sidereal time anywhere in the world. I am most grateful to all those people who took the trouble to write to me with their views and I hope that the process will continue for future printings of the book.

Some of the algorithms presented here are original. However, many of the subroutines are straightforward translations of the methods used by other authors. In particular, I have drawn heavily on Jean Meeus' excellent book *Astronomical Formulae for Calculators*, and I am much indebted to him for his clarity of style and presentation. The theory behind the calculations is not dealt with in this book since it is usually not of immediate interest when a particular problem has to be calculated.

My aim has been to provide a book which, above all else, is useful. If you persevere through the tedium of typing these lines of BASIC into your computer,† I hope you will find here a doorway into the realms of astronomical prediction which, I suppose, is limited only by your imagination.

Peter Duffett-Smith
Downing College, Cambridge
January 1989

---

† The programs are also available on floppy disk for some computers. Please see the back of the book for details.

USING YOUR PERSONAL COMPUTER FOR ASTRONOMY

Do you have your own computer? And do you wish to use it for predicting astronomical phenomena such as the time of moonrise or the circumstances of the next eclipse? If so, this book may help you. It provides a kit of parts, a set of building blocks, to write complex programs in astronomy tailored to your own needs. Each building block is in the form of a subroutine, written in BASIC, which performs a well-defined function. There are 33 such routines in all which can link together in any combination. They cover most of the problems likely to be tackled by the amateur astronomer, ranging from very simple tasks like converting between decimal hours and hours, minutes, and seconds, to complex procedures for deducing the set of orbital elements which is consistent with the observed movement of a minor planet or comet through the heavens. It is possible that your particular requirement may be covered directly by a subroutine as it stands, for example, the calculation of the position of the Moon. You can then enter it from the keyboard or copy it from disk† and run it with its supplied handling program. Or it may be that you have a much more demanding task, needing the results of several routines together to get the answer. Suppose, for instance, that you wished to predict the time of rising of a newly-discovered comet in Hawaii on the 27th August this year. You would then write a simple handling program, calling the subroutines supplied here, first to find the position of the comet in ecliptic coordinates, then to convert to equatorial coordinates, and finally to calculate the time of rising for an object in the sky at that right ascension and declination. You might have considered the task too daunting to attempt from scratch, but with most of the work already done by these routines, you'll soon find yourself contemplating ambitious projects and discovering the great satisfaction of predicting accurately astronomical events from your armchair (which you can later verify by drawing back the curtains!).

As I have already indicated, the routines are designed to be supervised by a handling program which asks the operator for the required parameters (e.g. the time and place), converts them into their correct formats (usually radians), calls

† See footnote on p. x.

1

the subroutines to make the calculations, and finally displays the results having first converted them into a readable format (e.g. from radians to degrees, minutes, and seconds). I have supplied a handling program with every sub-routine in the book, and in some cases I have included extra ones where they are particularly useful. The handling programs begin with 'H'; thus subroutine MINSEC can be run by using the handling program HMINSEC, and ELOSC by HELOSC etc. Other particularly useful handling programs have names describing their functions, such as HALTAZ which calculates the *al*titude and *az*imuth of a heavenly body given its equatorial coordinates, or HSUNEP which displays an *ep*hemeris of the *Sun*. You'll find a table of all these handling programs later in this section (Table 2).

One of the problems with a book of this sort comes about because there is such a wide variety of computers available. Programs written for one machine will probably not work properly on another, if at all. BASIC is supposed to be a universal computer language, but in practice there are significant differences between the exact syntax used here and the intrinsic functions available there. Then again, even supposing the program itself runs properly, the method by which the results are presented, in graphical form or otherwise, is likely to be entirely different from machine to machine. I'm afraid those programs will only work well on the machines for which they were written, and there is often insufficient detail supplied for making the transfer easily to another model.

I have attempted to overcome this problem in several ways. First, I have used only a subset of BASIC which is common to nearly every machine I have met. Where there are slight differences, the program structure is always such as to allow easy changes. For example, the command ON . . . GOSUB can easily be replaced by several IF . . . THEN statements. I have avoided using functions which are not universally supported and have kept the listing of results by the handling programs as simple as possible. You can adapt these to suit yourself. Second, I have not attempted to present dazzling graphics displays such as you see in video games. Although such displays greatly enhance the satisfaction of the user and are most efficient in presenting the data in intelligible form, they are heavily 'machine-dependent' making it virtually impossible to write sensible universal code. Where a graphics display is especially important, as in the presentation of an eclipse, I have listed the routine which works on my machine as a basis for modification with the help of detailed notes and instruc-tions. In any case, I feel that you will get much greater satisfaction from these programs if you write important parts yourself. In general then, I solve the astronomy while you solve the problem of displaying the results. Third, the program listings are annotated to help make it clear what is going on. The BASIC statements appear on the right-hand side of the page, while on the left-hand side are written notes opposite the statements to which they apply. There is ample space here, too, for you to add your own comments if you wish. When read in conjunction with the formulae and the brief introductory remarks of each section, I hope that it will be easy to see how every routine works so that you can modify it as necessary.

The astronomy subroutines (see Table 1) begin at line number 1000, each occupying a different area so that all could be used in the same program without overlap. I have also included two extra routines to help with the writing of the handling programs. These begin at 880 and 960. The eclipse display routine begins at line 500. Other than this, all the space from line number 1 to 999 is available for the handling routine. Line numbers generally increment by 5 from one to the next, leaving room for expansion. This is quite important as you may wish to made modifications, either to calculate different results or because you need to get around a particular peculiarity of your machine. Table 1 lists the routines by name, the line numbers at which they begin (generally their entry points), and the various flags, error codes, and direction switches associated with them. The flags are generally used to indicate whether a routine has been called before. Often there is a chunk of code which needs to be executed once only to calculate quantities which do not change from call to call. The flag causes execution to by-pass such code after the first call to the routine, thus saving time. The error codes indicate problems which arise during the running of the program. For example, ER(3) (usually set to 0) is set to 1 if the celestial object in question never rises. The direction switches specify which direction conversions are to be carried out. With SW(3) equal to 1, for example, the input parameters are converted from equatorial to ecliptic coordinates, and vice-versa if SW(3) $= -1$.

All the programs listed in this book have been tested using both a double-precision version of BASIC and a single-precision version. There are some differences in presentation of output, but otherwise the code runs under both interpreters without modification. The examples listed in this book are the results of running the programs with the double-precision version. This represents each real number using 8 bytes to achieve better than 12-digit accuracy. The other version, however, uses 4 bytes and achieves only 6 or 7-digit accuracy. You should check with your user manual to discover the accuracy of your particular machine. Occasionally, the lower accuracy of single precision is not sufficient, as for example when extrapolating far into the future or the past. Compare your results with the examples listed here. You should get reasonable agreement, but make allowance for the rounding errors unavoidable if your machine uses a lower accuracy than mine.

Many of the routines call other routines in the course of their calculations.

Using your personal computer for astronomy

Table 1

Subroutine name	Line number	Flags and switches		
		Flag	Error	Switch
DISPLAY	500	—	—	—
DEFAULT	880	—	—	—
YESNO	960	—	—	—
MINSEC	1000	—	—	SW(1)
JULDAY	1100	FL(1)	ER(1)	—
CALDAY	1200	FL(2)	—	—
TIME	1300	FL(3)	ER(2)	SW(2)
EQHOR	1500	FL(4)	—	—
HRANG	1600	—	—	—
OBLIQ	1700	FL(5)	—	—
NUTAT	1800	FL(6)	—	—
EQECL	2000	FL(7)	—	SW(3)
EQGAL	2100	FL(8)	—	SW(4)
GENCON	2200	FL(4),FL(7),FL(9)	—	—
PRCESS1	2500	FL(10)	—	—
PRCESS2	2600	FL(11)	—	—
PARALLX	2800	FL(12)	—	SW(5)
REFRACT	3000	—	—	SW(6)
RISET	3100	FL(4)	ER(3)	—
ANOMALY	3300	—	—	—
SUN	3400	—	—	—
SUNRS	3600	—	ER(4),ER(5)	SW(7)
PELMENT	3800	—	—	—
PLANS	4500	—	—	—
MOON	6000	—	—	—
MOONRS	6600	—	ER(6),ER(7)	—
MOONNF	6800	—	—	—
ECLIPSE	7000	—	ER(8),ER(9)	—
ELOSC	7500	—	—	—
RELEM	7700	—	—	—
PCOMET	7900	—	—	—
PFIT	8100	—	ER(10)	—
EFIT	8800	—	ER(11)	—

Table 2

Handling program						

```
                DISPLAY
                  DEFAULT
                    YESNO
                      MINSEC
HMINSEC      — — C C  JULDAY
HJULDAY      — — C — C  CALDAY
HCALDAY      — — C C — C
HCALEND      — — C C C C  TIME
HTIME        — — C C C — C  EQHOR
HEQHOR       — — C C — — — — C  HRANG
HHRANG       — — C C C — C — C  OBLIQ
HALTAZ       — — C C C — C C    NUTAT
HOBLIQ       — — C C C — — — — C C
HNUTAT       — — C C C — — — — — C  EQECL
HEQECL       — — C C C — — — — — C C C  EQGAL
HEQGAL       — — C C — — — — — — — — C  GENCON              Subroutines
HGENCON      — — C C C — C — — C C — — — C  PRCESS1
HPRCESS1     — — C C C — — — — — — — — — — C  PRCESS2
HPRCESS2     — — C C C — — — — — — — — — — — C  PARALLX
HPARALLX     — — C C C — C — C — — — — — — — — C  REFRACT
HREFRACT     — — C C C C C — — — — — — — — — — C  RISET
HRISET       — — C C C — C — — — — — — — — — — — C  ANOMALY
HANOMALY     — — C C — — — — — — — — — — — — — — — C  SUN
HSUN         — — C C C — — — — C C C — — — — — — — C C
HSUNEP       — — C C C C — — — — C C C — — — — — — C C  SUNRS
HSUNRS       — — C C C — C — — — C C C — — — — — — C C C C  PELMENT
HPELMENT     — — C C C — — — — — — — — — — — — — — — C  PLANS
HPLANS       — — C C C — — — — C C C — — — — — — — C C — C C
HPLANEP      — — C C C C — — — — C C C — — — — — — C C — C C
HPLANRS      — — C C C — C — — — C C C — — — — — — C C C — C C  MOON
HMOON        — — C C C — — — — C C C — — — — — — — C C — — — C
HMOONEP      — — C C C C — — — — C C C — — — — — — C C — — — C  MOONRS
HMOONRS      — — C C C — C — — — C C C — — — — — — C — — — — — C C  MOONNF
HMOONNF      — — C C C C — — — — — — — — — — — — — — — — — — — — C  ECLIPSE
HECLIPSE     C — C C C C C — C C C C — C — — — C — — — C C — — — C — C C  ELOSC
                                                                           RELEM
HELOSC       — C C C C — — — — C C C — — C — — — — — C C — — — — — — — C C
HRELEM       — — C — C — — — — — — — — — — — — — — — — — — — — — — — — — — C  PCOMET
HPCOMET      — C C C C — — — — C C C — — C — — — — — C C — — — — — — — — — C C  PFIT
HPFIT        — C C — C C — — — — — — — — — — — — — — C C — — — — — — — — — — — C  EFIT
HEFIT        — C C — C C — — — — — — — — — — — — — — C C — — — — — — — — — — — — C
```

Table 2 indicates the routines required by each of the handling programs named on the left-hand side of the table. You must include them all before running the program.

This book is divided into separate sections, each dealing with one or two subroutines. Each section begins with a few introductory paragraphs describing the routines and the astronomical context into which they fit. I have assumed a working knowledge of basic astronomy and have not attempted to give elaborate explanations of the astronomical phenomena. Next there are notes on the formulae used by the routines (where appropriate) and then a condensed set of instructions noting the variables, error flags, etc. This is followed by listings of the subroutines themselves and then one or more handling programs, the latter designed to show how the routines should be used. Note that when you enter the code into your computer, you must include *all* the subroutines called in the course of the calculation. Thus HECLIPSE will require you to enter more than a dozen separate routines. Finally, there is an example showing the result of running each handling program with all output directed to the printer. This serves to illustrate the operation of the program and also allows you to check your results against mine to ensure that you have not made any errors in typing the code.

In this edition I have confined myself to using names for the variables which are only one or two characters long. This was necessary as there are still computers about which do not recognise names longer than these. You will find a list at the back of the book containing all the variables used in each of the routines in alphabetical order, so that when you make modifications or include new routines of your own, you can quickly determine which names have already been allocated. Alternatively, if your machine allows it, you can use longer names with no possibility of interference with the variables of my routines.

880 DEFAULT
960 YESNO

Routine DEFAULT allows for input from the keyboard with default settings. Routine YESNO handles input from the keyboard in response to questions requiring the answer 'yes' or 'no'.

DEFAULT

The handling programs in this book, whose names all begin with the letter H (e.g. HEQECL, HALTAZ, etc.) require the user to supply answers to questions such as 'Today's date?' or 'Geographical latitude?'. Often it is convenient to make use of default settings for the answers, especially if they are many characters in length. Routine DEFAULT allows you to do this very simply. To use it, you supply the question in the string variable Q\$, the number of variables to be entered in N (= 1, 2, or 3) and the default settings for those variables in X, Y, and Z. The routine then displays the question on the screen together with the default settings in square brackets [ ], and waits for a response. You may enter new numbers for the variables as usual if you wish. Alternatively, you can enter a comma (,) in place of every variable, and the routine uses the default values instead. The new values, or default settings, are returned in the same variables X, Y, and Z.

Routine DEFAULT is used later in the book in the handling programs HELOSC, HPCOMET, HPFIT, and HEFIT. In HELOSC, for example, lines 160 and 165 get the date of the epoch (D,M,Y) by setting Q\$="Epoch (D,M,Y)", X, Y, and Z to the default values of the day, month, and year respectively, and N=3 (since there are three variables to assign). Suppose that X, Y, and Z were given the default values 12, 3, 1986 respectively. Then on calling the routine, you could respond with ',,,' (followed by the return or enter key) and control would return to the calling program with X, Y, and Z still set at their initial values. Alternatively, you might respond with ',,2003' to change Z to 2003. If you entered '19,4,1987' then X would be set to 19, Y to 4, and Z to 1987.

DEFAULT may also be used for string variables, as in lines 60 and 65 of HPCOMET. Q\$ carries the question as before, but set N=0 to signify that a string variable is required, and X\$ to its default value. X\$ is changed to the new string, if one is entered, or remains set to its default value is you respond with a comma.

880 DEFAULT and 960 YESNO

Some BASIC interpreters allow you to accept the default settings by simply pressing the return or enter key. You should experiment with your own machine to establish whether you may do so as well. Check to make sure that all the variables are correctly set to their default values, and not to zero!

Details of DEFAULT

Called by GOSUB 880.

If N=1, 2, or 3:

Accepts new values for variables X (N=1), X and Y (N=2), or X and Y and Z (N=3) in response to a question displayed on the screen which is carried by Q\$. The default values of the variables are also displayed in square brackets after the question.

If N=0:

Accepts a new or default string in X\$. Otherwise as for N=1.

YESNO

Routine YESNO handles questions which require a simple yes or no response. Once again the question itself is carried by the string variable Q\$ (which should indicate that 'Y' is expected for yes and 'N' for no). When the routine is called, it displays the question on the screen and waits for an answer. If it is 'Y' or 'y', the routine returns with the variable E set to 1. If the answer is 'N' or 'n', E is set to 0. Any other answer produces the response 'What?' and the question is repeated. YESNO is used by all the handling programs.

Details of YESNO

Called by GOSUB 960.

Displays the question carried by Q\$ on the screen and awaits a single-character response. It returns with the variable E=1 if the answer is 'Y' or 'y', or E=0 if the answer is 'N' or 'n'. Any other input causes 'What?' to be displayed followed by the question again.

```
877   REM
878   REM          Default value input routine
879   REM
```

Notes	
880	Get a string variable?
885	Convert default values to strings
890	Jump to appropriate input statement, depending on N
895	Input statements . . .
900	. . . for one variable
910	. . . for two variables
920	. . . for three variables
925	Assign default values if no new input
945	No conversion if string input (N=0)
950	Convert from strings to real numbers, and return

```
880   IF N=0 THEN D1$=X$ : GOTO 890
885   D1$=FNQ$(X) : D2$=FNQ$(Y) : D3$=FNQ$(Z)
890   ON N+1 GOTO 895,895,905,915
895   PRINT Q$+" ["+D1$+"] ";TAB(50);
900   INPUT A1$ : GOTO 925
905   PRINT Q$+" ["+D1$+","+D2$+"] ";TAB(50);
910   INPUT A1$,A2$ : GOTO 925
915   PRINT Q$+" ["+D1$+","+D2$+","+D3$+"] ";TAB(50);
920   INPUT A1$,A2$,A3$ : GOTO 925
925   IF A1$="" AND N=0 THEN A1$=D1$
930   IF A1$="" AND N>0 THEN A1$=STR$(X)
935   IF A2$="" THEN A2$=STR$(Y)
940   IF A3$="" THEN A3$=STR$(Z)
945   IF N=0 THEN X$=A1$ : RETURN
950   X=VAL(A1$) : Y=VAL(A2$) : Z=VAL(A3$) : RETURN
```

```
957   REM
958   REM          Input yes/no routine
959   REM
```

Notes	
960	Print question and get answer
965	Yes?
970	. . . or no?
975	Neither: go round again

```
960   PRINT Q$; : INPUT A$
965   IF A$="Y" OR A$="y" THEN E=1 : RETURN
970   IF A$="N" OR A$="n" THEN E=0 : RETURN
975   PRINT "What ? " : GOTO 960
```

9

1000 MINSEC

This routine converts decimal degrees/hours into degrees/hours, minutes, and seconds, and vice-versa.

The starting and finishing points of astronomical calculations are usually angles, expressed as degrees, minutes and seconds of arc, or times, measured in hours, minutes, and seconds. Yet computers often cannot handle these quantities as they stand. They must first be converted to decimal degrees or hours before the computation can begin, and then the result converted back into the more-familiar minutes and seconds form. Consider, for example, the problem of finding the time corresponding to 1 hour 37 minutes before 3 : 25 in the afternoon. We first convert the times to their decimal forms, 1.616667 hours and 3.416667 hours, then subtract the former from the latter to get 1.800 hours, and finally convert back to minutes and seconds: 1 : 48 pm.

Your computer can deal with the problem of converting angles or times between decimal and minutes/seconds forms with the aid of the subroutine MINSEC. The direction in which the conversion is carried out is controlled by the switch SW(1). When SW(1) has the value $+1$, then the quantity represented by the input argument X (decimal degrees/hours) is converted into degrees/hours, minutes, and seconds which are returned by the output arguments XD, XM, XS respectively. XD and XM represent whole numbers while XS represents the seconds correct to two decimal places. The sign of X is returned by the string variable S$ which has the value '+' unless X is negative when S$ becomes '−'. The real variable SN also carries this information since SN $= +1$ if X is positive, SN $= 0$ if X is zero, and SN $= -1$ if X is negative.

Conversion in the other direction occurs if SW(1) $= -1$. Then the quantities XD (degrees/hours), XM (minutes), and XS (seconds) are converted into decimal form and returned in X. XD, XM, and XS must all be positive unless the angle or the time which they collectively represent is negative, in which case the negative sign can precede any non-zero element. For example, 137° 21′ 8″2 would be input as XD $= 137$, XM $= 21$, XS $= 8.2$, while -21° 46′ 3″ would be input as XD $= -21$, XM $= 46$, XS $= 3$, and -0° 15′ 4″ would be XD $=0$, XM $= -15$, XS $=4$.

MINSEC also outputs the result of the conversion in string form in the string variable OP\$. Thus it is possible to call the routine and display the result of the conversion in either direction with the statement PRINT OP\$. The number of characters in the string when $SW(1)=-1$ is set by the input parameter NC. If you forget to set it (leaving it equal to zero) then NC is set by the routine to 9. The number of decimal places displayed on the screen may be adjusted with NC.

MINSEC makes use of the BASIC function INT(X). In most personal computers, INT(X) is the least-integer function, returning the value of the nearest integer which is less than or equal to X. Thus $INT(21.7) = 21$ and $INT(-21.7) = -22$. However, some BASIC interpreters and compilers return the truncated-integer value of X, so that $INT(21.7) = 21$ as before, but $INT(-21.7) = -21$. It does not matter for this subroutine which form of INT(X) is used by your computer, although in later routines special measures have had to be adopted (see JULDAY and CALDAY).

Details of MINSEC

Called by GOSUB 1000.

If $SW(1) = +1$:

Converts X (decimal degrees/hours) into S\$ ('+' or '−'), XD (degrees/ hours), XM (minutes), XS (seconds) and SN $(1,0,-1$ if X is positive, zero, or negative respectively).

If $SW(1) = -1$:

Converts XD (degrees/hours), XM (minutes), XS (seconds) into X (decimal degrees/hours). Input the numbers with the negative sign (if any) preceding any non-zero element. The sign of X is also returned by the string variable S\$ ('+' or '−').

In both cases:

The result of the conversion is returned in the string variable OP\$. If $SW(1)=-1$ the number of characters in the string is determined by input parameter NC, set to 9 by the routine if it is 0 on entry. The string begins with the character '+' if X is positive, or '−' if X is negative.

```
                          997   REM
                          998   REM       Subroutine MINSEC
                          999   REM
```

1000 Which direction? Jump if minutes/
       seconds to decimal
1005 XP rounded to two decimal places in XS

1025 Form character string of the result . . .
1030 . . . in such a way that the minutes and
       seconds columns . . .
1035 . . . always line up with each other. This
       makes . . .
1040 . . . the output look neater.
1045 Return from MINSEC

1050 Minutes/seconds to decimal
1055 Any non-zero element carries the sign

1065 Set number of characters in OP$ if you've
       forgotten to
1070 Form character string of the result with
       NC characters
1075 Add sign if not already there
1080 Return from MINSEC

```
1000  IF SW(1)=-1 THEN GOTO 1050

1005  SN=SGN(X) : XP=ABS(X)+1.39E-6 : XD=INT(XP)
1010  A=(XP-XD)*60 : XM=INT(A) : S$="+"
1015  XS=INT((A-XM)*6000.0)/100.0
1020  IF SN=-1 THEN S$="-"
1025  A$=RIGHT$(("   "+STR$(XD)),3)
1030  B$=RIGHT$(("   "+STR$(XM)),3)

1035  D$=RIGHT$(("0"+STR$(INT((XS-INT(XS))*100))),2)

1040  C$=RIGHT$(("    "+STR$(INT(XS))+"."+D$),6)
1045  OP$="   "+S$+A$+B$+C$ : RETURN

1050  SN=+1 : S$="+"
1055  IF XD<0 OR XM<0 OR XS<0 THEN SN=-1 : S$="-"
1060  X=((((ABS(XS)/60)+ABS(XM))/60)+ABS(XD))*SN
1065  IF NC=0 THEN NC=9

1070  OP$=LEFT$(STR$(X),NC)

1075  IF SN=1 THEN OP$=S$+OP$
1080  OP$="   "+OP$ : RETURN
```

12

Notes	**HMINSEC**

```
1     REM
2     REM        Handling programme HMINSEC
3     REM

5     DIM SW(20)
10    PRINT : PRINT
15    PRINT "Decimal and minutes-seconds forms"
20    PRINT "--------------------------------"
```

Notes	Code
25 Set direction	`25    Q$="Convert to mins & secs (Y or N) "`
30 Call YESNO for answer	`30    PRINT : GOSUB 960`
	`35    IF E=1 THEN GOTO 80`
	`40    Q$="..... from mins & secs (Y or N) "`
	`45    GOSUB 960`
50 Jump if neither direction!	`50    IF E=0 THEN GOTO 105`
55 Minutes/seconds to decimal	`55    Q$="Degrees/hours, min, sec ....... "`
60 Get values . . .	`60    PRINT : PRINT Q$; : INPUT XD,XM,XS`
65 . . . call MINSEC to convert . . .	`65    SW(1)=-1 : NC=9 : GOSUB 1000`
	`70    Q$=".....  converts to ............. "`
75 . . . print the result	`75    PRINT Q$+OP$ : GOTO 105`
80 Decimal to minutes/seconds	`80    Q$="Decimal degrees/hours ......... "`
85 Get value . . .	`85    PRINT : PRINT Q$; : INPUT X`
90 . . . call MINSEC to convert . . .	`90    SW(1)=1 : GOSUB 1000`
	`95    Q$=".....  converts to ............. "`
100 . . . print the result	`100   PRINT Q$+OP$`
105 Another conversion?	`105   Q$="Again (Y or N) ............... "`
110 Call YESNO to find out	`110   PRINT : GOSUB 960`
	`115   IF E=1 THEN GOTO 25`
	`120   STOP`

```
INCLUDE YESNO, MINSEC
```

1000 MINSEC

Example

```
Decimal and minutes-seconds forms
----------------------------------

Convert to mins & secs (Y or N) ? Y

Decimal degrees/hours ........ ? 238.91945
..... converts to ...........   +238 55 10.02

Again (Y or N) ............... ? Y

Convert to mins & secs (Y or N) ? N
..... from mins & secs (Y or N) ? Y

Degrees/hours, min, sec ....... ? 238,55,10.02
..... converts to ...........   +238.91945

Again (Y or N) ............... ? Y

Convert to mins & secs (Y or N) ? Y

Decimal degrees/hours ........ ? -0.02234
..... converts to ...........   -  0  1 20.42

Again (Y or N) ............... ? Y

Convert to mins & secs (Y or N) ? N
..... from mins & secs (Y or N) ? Y

Degrees/hours, min, sec ....... ? 0,-1,20.42
..... converts to ...........   -0.022338

Again (Y or N) ............... ? N
```

1100 JULDAY

This routine calculates the number of days elapsed since the epoch 1900 January 0.5 (i.e. noon at Greenwich on 31st December 1899)

Many problems in computational astronomy take a date and a time as their starting points. For example, we may wish to calculate the position of Venus at 4 : 30 pm on 17th August 1938, or the phase of the Moon at midnight on the 19th May 1997. These dates and times are reckoned according to some nationally-agreed calendar which ascribes numbers to each instant so that it may be identified uniquely. If we say that a particular event occurred at 11 o'clock in the morning on 17th August 1938, we usually mean that it occurred at 11 hours after the beginning of the 17th day after the beginning of the 8th month after the beginning of the 1938th year after the adopted instant of the birth of Christ. At least, this is what we mean if we are using the Gregorian calendar, which is usually the case.

In order that we may make use of the given date in a computer program, it must first be converted to a single number which can be handled easily by the machine. We must choose some particular instant as our starting point, and then count the days logically from that moment. There are several obvious choices. We could, for example, choose the agreed instant of the birth of Christ, as the Gregorian calendar seems to do. We have to remember to count in the correct number of days for each month (28, 29, 30, or 31) and then add them to the day number of the date. However, we come to a complication when we wish to add in the year as well. Before 5th October 1582, the Julian calendar was in general use in Europe. By this calendar there were 365 days in each year except if the year number was divisible by 4, when there were 366. This variation was incorporated to make some allowance for the fact that the length of the year, i.e. the time taken by the Earth to complete one orbit around the Sun, is not a whole number of days. The length of the tropical year is in fact 365.2422 days, and by adopting the convention of a leap year every fourth year, the average Julian year is 365.25 days. This worked quite well for hundreds of years, until by 1582 astronomers were worried by the accumulated error which had built up through the difference between the tropical year and the Julian year. Accordingly, Pope Gregory decreed that the dates 5–14th October 1582 inclusive were to be abolished (to readjust the seasons to their correct positions) and that years ending in two zeros (1700, 1800 etc.) were only leap years if divisible by 400. By this device, 400 civil years contained $(400 \times 365) + 100 - 3 = 146\,097$ days. The average length of the civil year was then $146\,097/400 = 365.2425$ days, a much better approximation to the length of the tropical year.

Pope Gregory's calendar is the one usually used today, and our conversion of the calendar date into the number of days elapsed since a given moment has to take account of the complications mentioned above. In fact, astronomers generally adopt the instant of midday as measured at Greenwich on 1st January of the year 4713 BC as their starting point, or fundamental epoch. Any given calendar date is then converted into the number of days elapsed since then and called the Julian day number, or Julian date. Thus 17th August 1938 becomes $JD = 2\,429\,127.5$. Note that Julian days begin at Greenwich (mean) noon, i.e. at 12 : 00. Hence the '.5' in the Julian date. We can add in the time as well by converting it to a fraction of a day and adding it to the Julian date. Thus 11 am becomes $11/24 = 0.4583$ days and 11 am on the 17th August 1938 is then $JD = 2\,429\,127.9583$.

We now run into another complication. Many personal computers cannot handle a number with 11 decimal digits in it without loss of accuracy. In some forms of BASIC, numbers are represented in the machine by four bytes, or 32 bits, giving a precision of six or seven digits only. Other machines use five bytes to increase the precision to eight or nine digits. The most that you are likely to encounter is 'double-precision' representation by eight bytes, with up to 16-digit accuracy. The examples given in this book were obtained using a version of BASIC with eight-byte precision. Most versions have a lower precision than this, and in this book I have generally assumed that you are using a four-byte representation. This is sufficient for day numbers of six digits before the decimal point and one after it. To ease the problem of converting dates to day numbers, I have chosen the epoch 1900 January 0.5 as the starting point. This greatly reduces the size of the day number for any date within a few centuries either side of the epoch, sufficiently that four-byte representation is enough. For large extrapolations into the future or the past, you must use a form of BASIC with greater precision.

Subroutine JULDAY takes the calendar date input as DY (day number, including time as a fraction of the day if appropriate), MN (month number) and YR (year number), and converts it to the number of days since 1900 January 0.5 returned by DJ. Thus 17th August 1938 is $DJ = 14107.5$. To convert DJ to JD, simply add $2\,415\,020$ (if you have sufficient precision). This routine works for all dates since 1st January 4713 BC (again, if the machine has sufficient precision).

The term 'before Christ', or BC for short, usually refers to the chronological system of reckoning negative years. In this system, there is no year zero. The Christian era begins with the year 1 AD ('*anno domini*'); the year immediately preceding it is designated 1 BC, and the year before that 2 BC. Astronomers, on the other hand, need to count the years logically, taking the year before 1 AD as 0, and the year before that as −1 etc. You need not be worried by all this, since subroutine JULDAY takes care of it for you. It assumes the chronological sequence of years BC −3, −2, −1, 1, 2, 3, AD, correcting for the absence of year 0 in line 1115. Input years BC simply with a negative sign. If you attempt to input YR=0 the routine responds with the message '** *no year zero* **', followed by '*** *impossible date*'. The latter message is also displayed if you attempt to input any date between 5th and 14th October 1582 inclusive. In both cases, the error flag ER(1) is set to 1 to indicate to the calling program that an error has occurred. Dates before 1900 January 0.5 return negative values of DJ.

Execution of the subroutine JULDAY is controlled by the flag FL(1). When it is called for the first time, or whenever it is required to calculate DJ for a new date, FL(1) must be set to 0. In line 1110 the flag is set to 1. If, subsequently, the routine is called again without the flag having been reset to 0, DJ is *not* recalculated but control is immediately returned to the calling program (line 1100). This is to avoid recalculating DJ for the same date when stringing together several routines into one program, each of which makes a call to JULDAY.

Subroutine JULDAY uses the BASIC function INT(X) to find the integer part of the number represented by X. Unfortunately, some forms of BASIC use INT to represent the least-integer function while others use it to represent the truncated-integer value. Thus INT(−32.4) would be returned as −33 in the former case and as −32 in the latter. Both return INT of positive numbers identically. In order to overcome this difficulty, JULDAY uses its own integer function FNI(W), which returns the truncated integer part of W. This must be defined in the handling program.

Formulae

$DJ = B + C + D + DY - 0.5$ days

If date is before 15th October 1582

$B = 0$

If date is equal to, or after 15th October 1582

$B = 2 - A + INT(A/4)$

$A = INT(Y1/100)$

If Y1 is negative:

$C = FNI((365.25 \times Y1) - 0.75) - 694\,025$

1100 JULDAY

If Y1 is zero or positive:
$$C = INT(365.25 \times Y1) - 694025$$
All dates:
$$D = INT(30.6001 \times (M1+1))$$
If YR is negative:
$$Y1 = YR + 1$$
If YR is positive:
$$Y1 = YR$$
If MN = 1 or 2:
$$Y1 = Y1 - 1$$
$$M1 = MN + 12$$
If MN is 3 or more:
$$M1 = MN$$
INT( ) can return either the least-integer or the truncated-integer value. FNI( ) must return the truncated-integer value.

Details of JULDAY

Called by GOSUB 1100.

Converts the date DY (days), MN (months), YR (years) into DJ (days), the number of Julian days elapsed since 1900 January 0.5 (= 1899 December 31.5). Count years 'before Christ' as negative, with no year 0, thus: BC -3, -2, -1, 1, 2, 3 AD.

If FL(1) = 0 on call, execution of the subroutine proceeds normally. If FL(1) = 1 on call, then control returns immediately to the calling program without calculation of a new DJ. FL(1) is set to 1 on each new calculation.

ER(1) is set to 0 on return from the subroutine if no error has occurred. If the date is impossible (see text) then a message is displayed, ER(1) is set to 1, and the value of DJ is not altered.

```
                                    1097 REM
                                    1098 REM        Subroutine JULDAY
                                    1099 REM
```

1100 Has this date already been converted?
```
                                    1100 IF FL(1)=1 THEN RETURN
```

1105 Check that the year is legal
```
                                    1105 IF YR=0 THEN PRINT "** no year zero **" : GOTO 1150
```

```
                                    1110 M1=MN : Y1=YR : FL(1)=1 : B=0 : ER(1)=0
```
1115 No year 0
```
                                    1115 IF Y1<1 THEN Y1=Y1+1
                                    1120 IF MN<3 THEN M1=MN+12 : Y1=Y1-1
```
1125 Deal with change to Gregorian calendar
```
                                    1125 IF Y1>1582 THEN GOTO 1160
                                    1130 IF Y1<1582 THEN GOTO 1165
                                    1135 IF Y1=1582 AND MN<10 THEN GOTO 1165
                                    1140 IF Y1=1582 AND MN=10 AND DY<5 THEN GOTO 1165
                                    1145 IF MN>10 OR DY>=15 THEN GOTO 1160
```
1150 Jump here if illegal date
1155 Error return from JULDAY
```
                                    1150 ER(1)=1 : FL(1)=0
                                    1155 PRINT "*** impossible date" : RETURN
```

1160 Calculate A, B, C, and D
```
                                    1160 A=INT(Y1/100) : B=2-A+INT(A/4)
                                    1165 C=INT(365.25*Y1)-694025
                                    1170 IF Y1<0 THEN C=FNI((365.25*Y1)-0.75)-694025
                                    1175 D=INT(30.6001*(M1+1)) : DJ=B+C+D+DY-0.5
```
1180 Normal return from JULDAY
```
                                    1180 RETURN
```

Notes		
	1	REM
	2	REM Handling program HJULDAY
	3	REM
10 Define truncated-integer function FNI	5	DIM FL(20),ER(20)
	10	DEF FNI(W)=SGN(W)*INT(ABS(W))
	15	PRINT : PRINT
	20	PRINT "Calendar date to Julian days"
	25	PRINT "--------------------------"
30 Set FL(1)=0 for new date	30	PRINT : FL(1)=0
	35	Q$="Calendar date (D,M,Y; BC negative) .. "
40 Get date DY,MN,YR	40	PRINT Q$; : INPUT DY,MN,YR
45 Call JULDAY for DJ, and calculate JD	45	GOSUB 1100 : JD=DJ+2415020.0
50 ER(1) non-zero if illegal date	50	IF ER(1)=1 THEN GOTO 70
	55	Q1$="J days since 1900 Jan 0.5 .... "
	60	Q2$="Julian date......................... "
65 Display the results neatly	65	PRINT Q1$+" "+STR$(DJ) : PRINT Q2$+" "+STR$(JD)
70 Another conversion?	70	Q$="Again (Y or N) .... "
75 Call YESNO to find out	75	PRINT : GOSUB 960
	80	IF E=1 THEN GOTO 30
	85	STOP

```
INCLUDE YESNO, JULDAY
```

Example

```
Calendar date to Julian days
----------------------------

Calendar date (D,M,Y; BC negative) .. ? 2,2,1989
J days since 1900 Jan 0.5 ..........   32539.5
Julian date.........................   2447559.5

Again (Y or N) ..................... ? Y

Calendar date (D,M,Y; BC negative) .. ? 0.5,1,1900
J days since 1900 Jan 0.5 ..........   0
Julian date.........................   2415020

Again (Y or N) ..................... ? Y

Calendar date (D,M,Y; BC negative) .. ? 10,10,1582
*** impossible date

Again (Y or N) ..................... ? Y

Calendar date (D,M,Y; BC negative) .. ? 23,11,0
** no year zero **
*** impossible date

Again (Y or N) ..................... ? Y

Calendar date (D,M,Y; BC negative) .. ? 1.5,1,-4713
J days since 1900 Jan 0.5 ..........   -2415020
Julian date.........................   0

Again (Y or N) ..................... ? N
```

1200 CALDAY

This routine converts the number of Julian days since 1900 January 0.5 into the calendar date (the inverse function of JULDAY).

The result of an astronomical calculation is sometimes an instant of time, expressed as a number of days since a fundamental epoch. In our convention, that epoch is 1900 January 0.5, and the number of days elapsed since then is represented by the real variable DJ. For example, we may have calculated the time of the next lunar eclipse as DJ = 30815.69, but unless we are familiar with the interpretation of such numbers, the result will not mean very much to us. We need a conversion routine to convert a Julian day number into a calendar date, the inverse function of JULDAY.

Subroutine CALDAY performs this function. It converts DJ, the number of Julian days (and fraction of a day) elapsed since 1900 January 0.5, into the calendar date represented by DY, the day of the month including the fraction of the day, MN, the month number, and YR, the year number. The fraction of the day is also returned by FD. Thus DJ = 30815.69 gives DY = 15.19, MN = 5, YR = 1984 and FD = 0.19, from which we conclude that the lunar eclipse was in progress on 1984 15th May at 4 : 33 UT (Universal Time). The integer part of the day number is returned in the variable ID, and a string representation of the entire date in the string variable DT$. Thus it is possible to call the routine and display the result immediately using the statement PRINT DT$.

CALDAY is valid at least for all DJ numbers greater than or equal to −2415020 (JD = 0) corresponding to noon on 1st January 4713 BC. Years 'before Christ' (BC) are counted as negative with no year 0. Thus the sequence is BC −3, −2, −1, 1, 2, 3 AD (see JULDAY). Once again, the full accuracy of this routine depends upon the precision with which numbers are represented in your machine. If you are using four-byte representation, you can expect at most seven decimal digits to be significant and if you wish to work back as far as DJ = −2415020, you must expect an error of at least 1 day in the result. For higher accuracy, you must use a form of BASIC (or other computer language) which allows greater precision.

CALDAY makes use of the least-integer function to find the nearest integer whose value is equal to or less than X. Many personal computers represent this

1200 CALDAY

function as INT(X), but some use INT(X) for the truncated-integer value. To overcome this problem, CALDAY uses its own least-integer function FNL(W), which must be defined by the handling program. Execution of the routine is controlled by the flag FL(2). This flag must be set to 0 by the calling program when calling the routine for the first time, or whenever a new value of DJ is to be converted. The routine itself sets the flag to 1 each time it makes a calculation (line 1270). If, subsequently, the routine is called again with FL(2) = 1, then control is returned immediately to the calling program (line 1200) without calculation of a new date. This is to avoid recalculating the date for the same DJ when stringing together several subroutines, each of which makes a call to CALDAY.

Formulae

$DY = C - FNL(30.6001 \times G) + FD$

If G is greater than 13.5:

$MN = G - 13$

If G is less than 13.5:

$MN = G - 1$

If MN is greater than 2.5:

$YR = B + 1899$

If MN is less than 2.5:

$YR = B + 1900$

If YR is 0 or negative:

$YR = YR - 1$

All dates:

$G = FNL(C/30.6001)$

$C = I - (FNL(365.25 \times B) + 7.50001E{-}1) + 416$

$B = FNL((I/365.25) + 8.02601E{-}1)$

$I = FNL(D)$

$D = DJ + 0.5$

$FD = D - I$

If FD is equal to 1:

$FD = 0$

$I = I + 1$

If I is greater than $-115\,860$:

$I = I + 1 + A - FNL(A/4)$

$A = FNL((I/36524.25) + 9.9835726E{-}1) + 14$

FNL( ) returns the least-integer value.

22

Details of CALDAY

Called GOSUB 1200.

Converts DJ, the number of Julian days elapsed since 1900 January 0.5 (including the fraction of a day) into the calendar date DY (days, including the fraction), MN (months), YR (years) and FD (fraction of a day). Years 'before Christ' are counted as negative with no year 0, thus: BC -3, -2, -1, 1, 2, 3 AD. The integer part of the day number is returned in ID, and a string representation of the date in DT$.

If FL(2) = 0 on call, execution of CALDAY proceeds normally. If FL(2) = 1 on call, control returns immediately to the calling program without calculation of a new date. FL(2) is set to 1 on each new calculation.

```
1197 REM
1198 REM        Subroutine CALDAY
1199 REM
```

Notes		
1200	Has this date already been converted?	`1200 IF FL(2)=1 THEN RETURN`
		`1210 D=DJ+0.5 : I=FNL(D) : FD=D-I`
1215	If time is 24 : 00 then increment day	`1215 IF FD=1 THEN FD=0 : I=I+1`
1220	Deal with Gregorian change	`1220 IF I <= -115860 THEN GOTO 1235`
1225	FNL is the least-integer function	`1225 A=FNL((I/36524.25)+9.9835726E-1)+14`
		`1230 I=I+1+A-FNL(A/4)`
		`1235 B=FNL((I/365.25)+8.02601E-1)`
		`1240 C=I-FNL((365.25*B)+7.50001E-1)+416`
		`1245 G=FNL(C/30.6001) : MN=G-1`
		`1250 DY=C-FNL(30.6001*G)+FD : YR=B+1899`
		`1255 IF G>13.5 THEN MN=G-13`
		`1260 IF MN<2.5 THEN YR=B+1900`
1265	No year 0!	`1265 IF YR<1 THEN YR=YR-1`
		`1270 ID=INT(DY) : FL(2)=1`
1275	Set DT$ to a string representation . . .	`1275 A$=RIGHT$((" "+STR$(ID)),2)`
1280	. . . of the calendar date, so that . . .	`1280 B$=RIGHT$(("   "+STR$(MN)),3)`
1285	. . . the results may be displayed easily . . .	`1285 C$=RIGHT$(("      "+STR$(YR)),6)`
1290	. . . and neatly. Return from CALDAY	`1290 DT$="   "+A$+B$+C$ : RETURN`

Notes		Program
| | | ```
1    REM
2    REM        Handling program HCALDAY
3    REM

5    DIM FL(20),SW(20)
``` |
| 10 | Define the truncated-integer function | ```
10   DEF FNI(W)=SGN(W)*INT(ABS(W))
``` |
| 15 | Define the least-integer function | ```
15   DEF FNL(W)=FNI(W)+FNI((SGN(W)-1.0)/2.0)
20   PRINT : PRINT
25   PRINT "Julian days to calendar date"
30   PRINT "----------------------------"
35   PRINT
``` |
| 40 | Set FL(2) to 0 for new date | ```
40   PRINT : FL(2)=0
``` |
| 45 | JD or modified JD? | ```
45   Q$="Convert full Julian date (Y/N) ....... "
``` |
| 50 | Call YESNO to see | ```
50   GOSUB 960
55   IF E=1 THEN GOTO 70
60   Q$="Julian days since 1900 January 0.5 ... "
``` |
| 65 | Get modified JD | ```
65   PRINT Q$; : INPUT DJ : GOTO 80

70   Q$="Julian date ......................... "
``` |
| 75 | Get full Julian date and convert to modified form | ```
75   PRINT Q$; : INPUT JD : DJ=JD-2415020
``` |
| 80 | Call CALDAY to convert | ```
80   GOSUB 1200

85   Q$="Calendar date (BC negative).......... "
``` |
| 90 | Display date, and convert time using MINSEC | ```
90   PRINT Q$+DT$ : X=FD*24 : SW(1)=1 : GOSUB 1000
95   Q$="... and time (H M S) ................ "
``` |
| 100 | Display time neatly | ```
100  PRINT Q$+" "+MID$(OP$,4,12)
``` |
| 105 | Another go? | ```
105  Q$="Again (Y or N) ...................... "
``` |
| 110 | Call YESNO to find out | ```
110  PRINT : GOSUB 960
115  IF E=1 THEN GOTO 40
120  STOP

INCLUDE YESNO, MINSEC, CALDAY
``` |
```

1200 CALDAY

Example

```
Julian days to calendar date
----------------------------

Convert full Julian date (Y/N) ....... ? Y
Julian date ........................ ? 2447559.5
Calendar date (BC negative)...........    2   2  1989
... and time (H M S) .................    0   0  0.00

Again (Y or N) ..................... ? Y

Convert full Julian date (Y/N) ....... ? N
Julian days since 1900 January 0.5 ... ? 32539.6671
Calendar date (BC negative)...........    2   2  1989
... and time (H M S) .................    4   0 37.44

Again (Y or N) ..................... ? Y

Convert full Julian date (Y/N) ....... ? Y
Julian date ........................ ? 0.75
Calendar date (BC negative)...........    2   1 -4713
... and time (H M S) .................    6   0  0.00

Again (Y or N) ..................... ? Y

Convert full Julian date (Y/N) ....... ? N
Julian days since 1900 January 0.5 ... ? 0.8
Calendar date (BC negative)...........    1   1  1900
... and time (H M S) .................    7  12  0.00

Again (Y or N) ..................... ? N
```

Here is a calendar program which uses the functions of both JULDAY and CALDAY. It allows you to convert from calendar dates to Julian dates, and vice-versa. It also determines the day of the week on which the entered date occurs. ☛

Notes	Line	Code
	1	REM
	2	REM Calendar programme HCALEND
	3	REM
5 Array N$ carries the names of the days	5	DIM N$(7),FL(20),ER(20),SW(20)
10 We need these special integer funcions . . .	10	DEF FNI(W)=SGN(W)*INT(ABS(W))
15 . . . for JULDAY and CALDAY.	15	DEF FNL(W)=FNI(W)+FNI((SGN(W)-1.0)/2.0)
20 FNR truncates after two decimal places	20	DEF FNR(W)=INT(100.0*W)/100.0
25 SET up the array N$	25	N$(1)="Sunday" : N$(2)="Monday" : N$(3)="Tuesday"
	30	N$(4)="Wednesday" : N$(5)="Thursday"
	35	N$(6)="Friday" : N$(7)="Saturday"
	40	PRINT : PRINT
	45	PRINT "Calendar programme"
	50	PRINT "-------------------" : PRINT
55 Both flags 0 for new date	55	PRINT : FL(1)=0 : FL(2)=0
60 Which way to convert?	60	Q$="Convert calendar date (Y or N) ... "
65 Call YESNO to see	65	GOSUB 960
	70	IF E=1 THEN GOTO 190
	75	Q$="Convert Julian date (Y or N) ..... "
	80	GOSUB 960
	85	IF E=0 THEN GOTO 230
90 Full JD or days since 1900 Jan 0.5?	90	Q$="Modified JD (Y or N) .... "
95 Call YESNO for the answer	95	GOSUB 960
	100	IF E=1 THEN GOTO 125
	105	PRINT
	110	Q$="Julian date .... "
115 Get Julian date . . .	115	PRINT Q$; : INPUT JD
120 . . . and calculate DJ	120	DJ=JD-2415020 : GOTO 145
	125	PRINT
	130	Q$="J days since 1899 Dec 31.5 .... "
135 Get DJ directly . . .	135	PRINT Q$; : INPUT DJ
140 . . . and calculate Julian date	140	JD=DJ+2415020
145 Call CALDAY to convert to calendar date	145	GOSUB 1200
	150	Q$="Calendar date is (BC neg.) .... "
155 Display the calendar date, and find the time	155	PRINT Q$+DT$: X=FD*24 : SW(1)=1 : GOSUB 1000
	160	Q$=".... and time (H M S) .... "
165 Display the time neatly	165	PRINT Q$+" "+MID$(OP$,4,12)
170 This finds the day of the week; index I	170	B=INT((JD+1.5)/7)*7 : I=INT(JD+1.5-B)+1
	175	IF I<1 THEN GOTO 230
	180	Q$="Day of the week .... "
185 Display the day of the week	185	PRINT Q$+" "+N$(I) : GOTO 230
	190	Q$="Calendar date (D,M,Y; BC neg.) ... "
195 Input the calendar date	195	PRINT : PRINT Q$; : INPUT DY,MN,YR
200 Call JULDAY to convert to modified Julian date	200	GOSUB 1100 : JD=DJ+2415020
205 Error flag set if date is impossible	205	IF ER(1)=1 THEN GOTO 230
	210	Q1$="J days since 1899 Dec 31.5 .... "
	215	Q2$="Julian date .... "
220 Display the two sorts of Julian date . . .	220	PRINT Q1$+" "+STR$(FNR(DJ))
225 . . . and then jump to find the day of the week	225	PRINT Q2$+" "+STR$(FNR(JD)) : GOTO 170
230 Another conversion?	230	Q$="Again (Y or N) .... "
235 YESNO will find out	235	PRINT : GOSUB 960
	240	IF E=1 THEN GOTO 55
	245	STOP

INCLUDE YESNO, MINSEC, JULDAY, CALDAY

1200 CALDAY

Example

```
Calendar programme
-------------------


Convert calendar date (Y or N) ... ? Y

Calendar date (D,M,Y; BC neg.) ... ? 19,5,1950
J days since 1899 Dec 31.5 .......   18400.5
Julian date .....................   2433420.5
Day of the week .................   Friday

Again (Y or N) ..................  ? Y

Convert calendar date (Y or N) ... ? N
Convert Julian date (Y or N) ..... ? Y
Modified JD (Y or N) ............. ? Y

J days since 1899 Dec 31.5 ....... ? 18400.75
Calendar date is (BC neg.) .......   19  5  1950
.... and time (H M S) ...........    6  0  0.00
Day of the week .................   Friday

Again (Y or N) ..................  ? Y

Convert calendar date (Y or N) ... ? N
Convert Julian date (Y or N) ..... ? Y
Modified JD (Y or N) ............. ? N

Julian date ..................... ? 2433420.45
Calendar date is (BC neg.) .......   18  5  1950
.... and time (H M S) ...........   22 48  0.00
Day of the week .................   Thursday

Again (Y or N) ..................  ? Y

Convert calendar date (Y or N) ... ? Y

Calendar date (D,M,Y; BC neg.) ... ? 1,11,0
** no year zero **
*** impossible date

Again (Y or N) ..................  ? N
```

1300 TIME

This routine converts a given local civil time into the local sidereal time and vice-versa, also calculating Universal Time and Greenwich sidereal time.

The basis of all civil timekeeping on the Earth is Universal time, UT. It is a measure of time which conforms, within a close approximation, to the mean daily motion of the Sun. Roughly speaking, the time is 12 : 00 noon UT when the Sun crosses the meridian at Greenwich, and 24 hours of UT is the time between two successive passages of the Sun across that meridian. However, the Sun is not really a very good timekeeper. Although the Earth spins about its polar axis at a nearly-constant rate (there are small irregular perturbations), the position of the Sun in the sky changes daily due to the orbit of the Earth about the Sun. If that orbit were perfectly circular, and if the Earth's axis were exactly perpendicular to the plane of the orbit (the ecliptic), then the daily passage of the Sun would be regular throughout the year. In reality, the Earth's orbit is slightly elliptical so that the speed of the Earth varies throughout the year, and the Earth's axis is tilted by about 23.5 degrees from the perpendicular to the plane of the ecliptic. Both these effects cause the time measured by a sundial, true solar time, to run fast during some parts of the year and slow during others by up to 16 minutes with respect to a uniform clock. The difference between the solar time and the uniform clock is called the 'equation of time', and you would need to know its value if you wished to tell the time accurately from a sundial.

To take account of the Sun's apparent aberrations from perfect timekeeping, we imagine a fictitious object, called the 'mean Sun', which moves at a constant rate along the equator as if we were in true circular orbit about it with the Earth's polar axis exactly perpendicular to the orbital plane. The position of the mean Sun is arranged to coincide with that of the true Sun each year at the (mean) vernal equinox. Time reckoned with respect to the mean Sun at Greenwich is sometimes known as Greenwich Mean Time (although that name is ambiguous), but is now more properly called Universal Time.

It is obviously quite sensible for civil time to be regulated more or less directly by the apparent daily motion of the Sun. Astronomers, however, are also interested in the motions of stars and other celestial bodies both inside and outside the Solar System. During the course of one tropical year, the Earth makes about 366.25 turns about its polar axis, so that the 'fixed heavens', defined by the background of stars at very great distances from us, appear to rotate this many times about the Earth. During the same time, the Earth makes one complete revolution about the Sun, moving in the same sense of rotation about the Sun as about its own polar axis. The result is that the Sun appears to rotate exactly one turn less about the Earth in one year than do the stars. There are therefore about 365.25 solar days in the year, and 366.25 star or sidereal days. Sidereal time is reckoned by the daily transit of a fixed point in space (fixed with respect to the distant stars), 24 hours of sidereal time elapsing between any two successive transits. The sidereal day is thus shorter than the solar day by nearly 4 minutes, and although the solar and sidereal times agree once a year, the difference between them grows systematically as the months pass in the sense that sidereal time runs faster than solar time. Sidereal time (ST) is used extensively by astronomers since it is the time kept by the stars, and we need routine TIME to convert between solar and sidereal time.

UT is now formally defined by a mathematical formula as a function of ST. Thus both UT and ST are determined from observations of the diurnal motions of stars. The time-scale determined directly from such observations is designated UT0 and is very slightly dependent on the place of observation. When corrected for the shift in longitude of the observing station because of an irregular polar motion, the time-scale UT1 is obtained. This is the scale usually referred to as Universal Time. Both UT1 and ST suffer from slight irregularities in the rotation of the Earth, and are therefore not strictly uniform time scales. With the advent of atomic clocks, however, timekeeping has become ever more precise. International atomic time, TAI, is the scale of time resulting from analyses by the Bureau International de l'Heure in Paris of atomic standards in many countries. Related to this is the Coordinated Universal Time, UTC. This differs from TAI by an integral number of seconds (about 20 in 1984) and it is maintained within 0.9 seconds of UT1 by introducing occasional leap seconds (usually at the end of June and December). UTC is the time broadcast by some national radio stations (the 'time pips') and by standard time transmission radio services; it forms the basis of legal time systems on the Earth.

The amateur astronomer need not be too concerned by all this complexity. For most purposes we can take UT = UT0 = UT1 = UTC = GMT without noticing the difference. If you do require very high precision you may also need to know about Ephemeris Time (ET), Terrestrial Dynamical Time (TDT) and Barycentric Dynamical Time (TDB), for which I refer you to *The Astronomical Almanac* and the *Explanatory Supplement* thereto.

Times may be converted quite easily from UT to Greenwich mean sidereal time (SG) since there exists a simple relationship between them. However, a complication does arise when converting from SG to UT because the sidereal day is shorter than the solar day (see Figure 1). Hence there is a small range of sidereal times which occurs twice on the same calendar date. The ambiguous period is from 23h 56m 04s UT to 0h 03m 56s UT, i.e. about 4 minutes either side of midnight. The routine given here correctly converts SG to UT in the period before midnight, but not in the period after midnight when the ambiguity must be resolved by other means.

The local sidereal time, LST, is calculated from SG and the geographical longitude by

$$LST = SG + (GL/15),$$

where GL is the longitude in degrees (West negative). SG is the local sidereal time for the meridian of Greenwich, longitude 0 degrees. It is formally defined as the hour angle of the vernal equinox. Local civil time (the time shown by ordinary clocks in any place) is usually a whole number of hours before or after UT, depending on the time zone. Thus, the local civil time in time zone +4 hours is equal to UT+4, unless daylight saving is in operation when you will need to make a further correction of an hour or two. Subroutine TIME handles the complexities of all this for you. It will convert your local civil time into your local sidereal time, or the other way around, calculating on the way UT and SG. The direction of conversion is controlled by the switch SW(2), being from civil time to sidereal time when SW(2) = 1, and from sidereal time to civil time when SW(2) = −1. Set the variable TM equal to your local time in hours (civil or sidereal) and specify the date by setting the values of DY (day), MN (month), YR (years). You also need to declare your geographical longitude in degrees in GL (W negative), your time zone in TZ (W negative), and the number of hours added for daylight saving time in DS. For example, to make the calculation on 22nd August 1988 in the UK when British Summer Time was in operation, on longitude 5° W, you would set DY=22, MN=8, YR=1988, GL=−5, TZ=0 and DS=1.

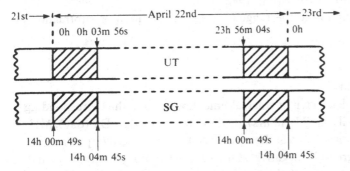

Figure 1. UT and SG for 22nd April 1980. The shaded intervals of SG occur twice on the same day.

1300 TIME

The results are returned in a variety of formats. The Universal Time in decimal hours is carried by the variable UT, and the hours, minutes, and seconds by UH, UM, and US respectively. There is also a string representation of the time available in UT$, so that you can display the result directly and neatly with the statement PRINT UT$. There are corresponding outputs for the Greenwich Sidereal Time in SG, GH, GM, GS, and SG$, and for the converted local time in TL, LH, LM, LS, and TL$. TIME makes calls to both MINSEC and JULDAY.

Execution of the subroutine is controlled by the flag FL(3). When the routine is called for the first time, or whenever a conversion is to be carried out for a new date, FL(3) must be set to 0. It is set to 1 by the routine. Subsequent calls to TIME on the same date need not have FL(3) = 0. The calculations from lines 1305 to 1330 inclusive depend only on the date and are skipped when FL(3) = 1, thus speeding up execution.

When the routine is used to convert from sidereal to civil time, there is a danger that the sidereal time lies in the ambiguous period described above. If this is the case, the routine displays the message '** *ambiguous conversion* **' and sets the error flag ER(2) equal to one. If there is no ambiguity, ER(2) is set to 0. The value of ER(2) is not changed if the conversion is from civil to sidereal time.

Formulae
The conversion between UT and SG uses the J2000.0 definition.

$SG = (UT \times 1.002737908) + T0$

$T0 = [R0 + R1]_{24}$

$R0 = (5.13366E{-}2 \times T) + (2.586222E{-}5 \times T^2) - (1.722E{-}9 \times T^3)$

$R1 = 6.697374558 + (2400 \times (T - ((YR - 2000)/100)))$

$T = (DJ/36525) - 1$

DJ = Integer Julian days since 1900 January 0.5

YR = Year number of date (e.g. 1992)

$[\ ]_{24}$ indicates reduction to the range 0–24 by the addition or subtraction of multiples of 24.

Details of TIME
Called GOSUB 1300.

Converts TM, the local civil or sidereal time (hours), into the corresponding local sidereal or civil time TL (hours), on the date specified by DY (day), MN (month), and YR (year), and at the place specified by the geographical longitude GL (degrees; W negative), time zone TZ (hours; W negative) and daylight saving correction DS (hours added). Also calculated are the Universal Time UT (hours) and Greenwich sidereal time SG (hours). TL, UT, and SG are also output in hours (LH, UH, GH), minutes (LM, UM, GM) and seconds (LS, US, GS) forms, and as strings in TL\$, UT\$, and SG\$ respectively.

Set SW(2) = +1 for conversion from civil to sidereal time.

Set SW(2) = −1 for conversion from sidereal to civil time.

Set FL(3) = 0 on first call of new date. FL(3) is set to 1 by TIME. Subsequent calls to TIME for the same date with FL(3) = 1 by-pass calculations which change only with the date, and hence speed up the process.

Error flag ER(2) is set to 0 by the routine if SW(2)= −1 and if the conversion is unambiguous. When an ambiguous conversion occurs, ER(2) is set to 1 and an error message is displayed. ER(2) is not affected if SW(2)=1.

Other routines called: MINSEC (1000), JULDAY (1100).

```
                               1297 REM
                               1298 REM        Subroutine TIME
                               1299 REM
```

1300	Called before for this date?	`1300 IF FL(3)=1 THEN GOTO 1335`
1305	Call JULDAY for JD of date	`1305 FL(3)=1 : FL(1)=0 : GOSUB 1100`
1310	Return if impossible date	`1310 IF ER(1)=1 THEN FL(3)=0 : RETURN`
1315	Calculate T0 for this date . . .	`1315 D=INT(DJ-0.5)+0.5 : T=(D/36525.0)-1`
		`1320 R0=T*(5.13366E-2+T*(2.586222E-5-T*1.722E-9))`
		`1325 R1=6.697374558+2400.0*(T-((YR-2000.0)/100.0))`
1330	. . . in range 0–24 hours	`1330 X=R0+R1 : GOSUB 1440 : T0=X`
1335	Which direction?	`1335 IF SW(2)=-1 THEN GOTO 1365`
1340	Civil to sidereal; correct for zone and saving	`1340 X=TM-DS-TZ : GOSUB 1440 : UT=X : T=T0`
1345	Allow for the date at Greenwich being different . . .	`1345 IF D=1 THEN T=T+6.57098244E-2`
1350	. . . from the local date	`1350 IF D=-1 THEN T=T-6.57098244E-2`
1355	Calculate SG in range 0–24	`1355 X=(UT*1.002737908)+T : GOSUB 1440 : SG=X`
		`1360 X=SG+(GL/15.0) : GOSUB 1440 : TL=X : GOTO 1400`
1365	Sidereal to civil; allow for zone and saving	`1365 X=T0-(DS+TZ)*1.002737908 : GOSUB 1440 : T=X`
1370	Find SG in range 0–24	`1370 X=TM-(GL/15.0) : GOSUB 1440 : SG=X : ER(2)=0`
1375	Allow for different dates	`1375 IF SG<T THEN X=X+24`
1380	Calculate civil time in range 0–24 . . .	`1380 X=(X-T)*9.972695677E-1 : GOSUB 1440 : TL=X`
1385	. . . and UT by correcting for zone and daylight saving	`1385 X=TL-DS-TZ : GOSUB 1440 : UT=X`
		`1390 Q$="** ambiguous conversion **"`
1395	Check if conversion is ambiguous	`1395 IF TL<6.552E-2 THEN ER(2)=1 : PRINT Q$`
1400	Use MINSEC to convert times to H,M,S forms . . .	`1400 X=UT : SW(1)=1 : GOSUB 1000`
1405	. . . UT . . .	`1405 UH=XD : UM=XM : US=XS`
1410	. . . and string representations	`1410 UT$=" "+MID$(OP$,4,12) : X=SG : GOSUB 1000`
1415	. . . SG . . .	`1415 GH=XD : GM=XM : GS=XS`
		`1420 SG$=" "+MID$(OP$,4,12) : X=TL : GOSUB 1000`
1425	. . . local time	`1425 LH=XD : LM=XM : LS=XS`
		`1430 TL$=" "+MID$(OP$,4,12)`
1435	Normal return from TIME	`1435 RETURN`
1440	Routine to return X in range 0–24 . . .	`1440 D=0`
1445	. . . and D signals change of day	`1445 IF X<0 THEN X=X+24 : D=-1 : GOTO 1445`
		`1450 IF X>24 THEN X=X-24 : D=1 : GOTO 1450`
		`1455 RETURN`

Notes		Code
	1	`REM`
	2	`REM        Handling program HTIME`
	3	`REM`
	5	`DIM FL(20),SW(20),ER(20)`
10 JULDAY needs this function	10	`DEF FNI(W)=SGN(W)*INT(ABS(W))`
	15	`PRINT : PRINT : SP=0 : FL(3)=0`
	20	`PRINT "Local time conversion"`
	25	`PRINT "--------------------" : PRINT : PRINT`
30 SP is a local switch to signify same place	30	`IF SP=1 THEN GOTO 70`
35 Get details of place . . .	35	`Q$="Daylight saving (H ahead of zone t)   "`
40 . . . daylight saving correction . . .	40	`PRINT Q$; : INPUT DS`
	45	`Q$="Time zone (hours; West negative) ... "`
50 . . . time zone	50	`PRINT Q$; : INPUT TZ`
	55	`Q$="Geog. longitude (D,M,S; W neg.) .... "`
60 . . . geographical longitude.	60	`PRINT Q$; : INPUT XD,XM,XS`
65 Call MINSEC to convert to decimal degrees	65	`SW(1)=-1 : GOSUB 1000 : GL=X`
70 FL(3)=1 if date has not changed	70	`IF FL(3)=1 THEN GOTO 85`
75 Otherwise, get a new one	75	`Q$="Calendar date (D,M,Y) ............. "`
	80	`PRINT : PRINT Q$; : INPUT DY,MN,YR`
85 Which direction to convert?	85	`Q$="Convert to sidereal time (Y or N) .. "`
90 YESNO finds out	90	`PRINT : GOSUB 960`
	95	`IF E=1 THEN SW(2)=1 : GOTO 120`
	100	`Q$="Convert from sidereal time (Y or N)  "`
	105	`GOSUB 960`
	110	`IF E=0 THEN GOTO 230`
	115	`SW(2)=-1 : GOTO 130`
120 Civil to sidereal . . .	120	`Q$="Local civil time (H,M,S) .......... "`
	125	`GOTO 135`
130 Sidereal to civil . . .	130	`Q$="Local sidereal time (H,M,S) ........ "`
135 Get local time . . .	135	`PRINT : PRINT Q$; : INPUT XD,XM,XS`
140 . . . and convert to decimal hours with MINSEC	140	`SW(1)=-1 : GOSUB 1000 : TM=X`
145 Call TIME . . .	145	`GOSUB 1300`
150 . . . and check for errors	150	`IF ER(1)=1 THEN GOTO 230`
155 Which way have we gone?	155	`IF SW(2)=1 THEN GOTO 200`
160 Display results for sidereal to local conversion	160	`Q$="Greenwich sidereal time (H,M,S) .... "`
	165	`PRINT Q$+SG$`
	170	`Q$="Universal time (H,M,S) ............ "`
	175	`PRINT Q$+UT$`
180 Extra error message if ambiguous conversion	180	`Q$="** may be in error by up to 4 minutes **"`
185 Print it if error flag was set to 1	185	`IF ER(2)=1 THEN PRINT Q$`
	190	`Q$="Local civil time (H,M,S) .......... "`
	195	`PRINT Q$+TL$ : GOTO 230`
200 Display results for local to sidereal conversion	200	`Q$="Universal time (H,M,S) ............ "`
	205	`PRINT Q$+UT$`
	210	`Q$="Greenwich sidereal time (H,M,S) .... "`
	215	`PRINT Q$+SG$`
	220	`Q$="Local sidereal time (H,M,S) ........ "`
	225	`PRINT Q$+TL$`

1300 TIME

```
230   Q$="Again (Y or N) .................... "
235   PRINT : GOSUB 960
240   IF E=0 THEN STOP
245   Q$="Same place (Y or N) .............. "
250   GOSUB 960 : SP=E
255   Q$="Same date (Y or N) ................ "
260   GOSUB 960 : FL(3)=E : PRINT : GOTO 30

      INCLUDE YESNO, MINSEC, JULDAY, TIME
```

Example

```
Local time conversion
---------------------


Daylight saving (H ahead of zone t) ? 1
Time zone (hours; West negative) ... ? -10
Geog. longitude (D,M,S; W neg.) .... ? -148,31,52.33

Calendar date (D,M,Y) .............. ? 11,3,1990

Convert to sidereal time (Y or N) .. ? Y

Local civil time (H,M,S) ........... ? 8,21,43.7
Universal time (H,M,S) .............   17 21 43.70
Greenwich sidereal time (H,M,S) ....    4 38  9.22
Local sidereal time (H,M,S) .......   18 44  1.73

Again (Y or N) .................... ? Y
Same place (Y or N) ................ ? Y
Same date (Y or N) ................. ? Y


Convert to sidereal time (Y or N) .. ? N
Convert from sidereal time (Y or N)   ? Y

Local sidereal time (H,M,S) ........ ? 18,44,1.73
Greenwich sidereal time (H,M,S) ....    4 38  9.22
Universal time (H,M,S) .............   17 21 43.70
Local civil time (H,M,S) ...........    8 21 43.70

Again (Y or N) .................... ? Y
Same place (Y or N) ................ ? Y
Same date (Y or N) ................. ? Y


Convert to sidereal time (Y or N) .. ? N
Convert from sidereal time (Y or N)   ? Y

Local sidereal time (H,M,S) ....... ? 10,23,0
** ambiguous conversion **
Greenwich sidereal time (H,M,S) ....   20 17  7.49
Universal time (H,M,S) .............    9  2  4.05
** may be in error by up to 4 minutes **
Local civil time (H,M,S) ...........    0  2  4.05

Again (Y or N) .................... ? N
```

1500 EQHOR

This routine converts celestial coordinates given in the horizon system (azimuth and altitude) into the corresponding equatorial coordinates (hour angle and declination), and vice-versa. (See also GENCON.)

A point in the sky may most easily be fixed by an observer on the Earth with reference to his horizon. In the horizon coordinate system (see Figure 2), the position of the point is specified by its *azimuth*, the angle round from the north† point of the horizon (in the sense NESW) and by its *altitude*, the angle up from the horizon (positive if above the horizon, negative if below it). The positions of heavenly bodies, on the other hand, are very often described in the equatorial coordinate system (see Figure 3). Here the plane of the Earth's equator, extended to cut the celestial sphere, is used instead of the horizon, with the *first point of Aires* or *vernal equinox* taking the place of the north point of the horizon. A star's position is then given by the angle round from the vernal equinox along the equator (in the opposite sense to that in which it appears to move throughout the day) and the angle up from the equator (positive if to the north, negative if to the south). These coordinates are called the *right ascension* and *declination* respectively. Related to the right ascension is the *hour angle* which describes the angle along the equator from the observer's meridian. As the Earth rotates, so the star and the vernal equinox appear to move at the same rate, making the right ascension and the declination constants (but see routines PRCESS1 and PRCESS2). The hour angle, however, increases uniformly throughout the day from zero when the star crosses the meridian (moving westerly). The right ascension and the hour angle are related by the formula

hour angle = LST − right ascension,

where LST is the local sidereal time (see TIME). Note that hour angle and right ascension are measured in opposite directions along the equator.

The routine given here converts between azimuth/altitude and hour angle/declination. The equations are symmetrical in the two pairs of coordinates so that exactly the same code may be used to convert in either direction, there being no need to specify the direction with a switch. You would use this

---

† Note that some authors use the south point of the horizon as reference.

routine, for example, to find where to look for the planet Mars in the sky having calculated its hour angle and declination.

The coordinates you wish to convert are input via the variables X (azimuth or hour angle) and Y (altitude or declination), and the results output via P (hour angle or azimuth) and Q (declination or altitude). You need also specify the geographical latitude, GP. All these angles must be expressed in radians. This use of X,Y for input and P,Q for output forms the general pattern in many of the subroutines given in this book.

EQHOR is controlled by the flag FL(4). This must be set to 0 when calling EQHOR for the first time or whenever a new geographical latitude is specified. It is set to 1 by the routine, and subsequent calls to the routine with FL(4) = 1 by-pass lines 1505 and 1510, hence saving time. Evaluation of a trigonometric function, sine, cosine, or tangent, and an inverse function, arcsine, arccosine, or arctangent, is a relatively-slow process. It is important to ensure that the sine of the same angle, for example, is not calculated more than once. The geographical latitude, GP, appears in the calculation as both COS(GP) and SIN(GP). These values are calculated once in line 1505 and then skipped unless GP changes (indicated by FL(4) = 0).

The handling program needs to define the special functions FNS and FNC used by the routine. These calculate values of arcsine and arccosine. If your computer already has these functions built into it, perhaps called ASIN and ACOS, you can replace FNS and FNC with the supplied functions if you so wish.

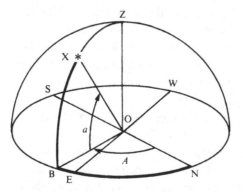

Figure 2. Horizon coordinates. The observer is at O. The plane of his horizon cuts the celestial sphere at NESW (North, East, South, and West points respectively). ZXB is the great circle through the zenith point Z and the star X and it cuts the horizon at B. The azimuth is the angle round from the N point (in the sense NESW) marked A, and the altitude is the angle up from the horizon marked a.

Formulae

$\cos(P) = (\sin(Y) - (\sin(GP) \times \sin(Q)))/(\cos(GP) \times \cos(Q))$

$\sin(Q) = (\sin(Y) \times \sin(GP)) + (\cos(Y) \times \cos(GP) \times \cos(X))$

P = hour angle or azimuth

Q = declination or altitude

given

X = azimuth or hour angle

Y = altitude or declination

GP = geographical latitude

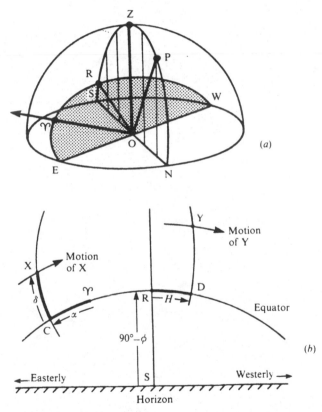

Figure 3. Equatorial coordinates: (a) on the celestial sphere, (b) as seen from the ground. NESW marks the plane of the observer's horizon, and Z his zenith point. ERW is the plane of the equator, and P the north pole. ♈ marks the direction of the vernal equinox. XC is an arc of the great circle through P and X, cutting the equator at C. α and δ are the right ascension and declination of X respectively. YD is an arc of the great circle through P and Y, cutting the equator at D. H is the hour angle of Y, and φ is the geographical latitude.

1500 EQHOR

Details of EQHOR

Called by GOSUB 1500.

Converts X, the azimuth or hour angle, and Y, the altitude or declination, into
P, the corresponding hour angle or azimuth, and Q, the corresponding declination or altitude, given GP, the geographical latitude (positive north). All
angles expressed in *radians*.

Set $FL(4) = 0$ on first call or for first calculation with new GP. $FL(4)$ is set to
1 by the routine and subsequent calls with $FL(4) = 1$ by-pass calculations
depending only on GP, hence saving execution time.

```
                              1497 REM
                              1498 REM        Subroutine EQHOR
                              1499 REM
```

1500	Previous call for this latitude?	`1500 IF FL(4)=1 THEN GOTO 1515`
1505	These values change only with latitude	`1505 CF=COS(GP) : SF=SIN(GP)`
1510	TP is 2π	`1510 FL(4)=1 : TP=6.283185308`
1515	Calculate trigonometric values once only	`1515 SY=SIN(Y) : CY=COS(Y)`
		`1520 SX=SIN(X) : CX=COS(X)`
		`1525 A=(SY*SF)+(CY*CF*CX)`
1530	FNS is arcsine	`1530 Q=FNS(A) : CQ=COS(Q) : B=CF*CQ`
1535	B is always positive; make sure it is never 0	`1535 IF B<1E-10 THEN B=1E-10`
1540	FNC is arccosine	`1540 CP=(SY-(SF*A))/B : P=FNC(CP)`
1545	Remove the ambiguity on taking arccosine	`1545 IF SX>0 THEN P=TP-P`
1550	Return from EQHOR	`1550 RETURN`

	1	REM
	2	REM Handling program HEQHOR
	3	REM
	5	DIM FL(20),SW(20)
10 FNM(W) converts W from degrees to radians	10	DEF FNM(W)=1.745329252E-2*W
15 FND(W) converts W from radians to degrees	15	DEF FND(W)=5.729577951E1*W
20 FNS(W) returns the inverse sine of W	20	DEF FNS(W)=ATN(W/(SQR(1-W*W)+1E-20))
25 FNC(W) returns the inverse cosine of W	25	DEF FNC(W)=1.570796327-FNS(W)
	30	PRINT : PRINT : FL(4)=0
	35	PRINT "Equatorial and horizon coords."
	40	PRINT "----------------------------"
	45	PRINT : PRINT
50 New latitude?	50	IF FL(4)=1 THEN GOTO 70
55 Yes – get value . . .	55	Q$="Geographical latitude (D,M,S; S neg.) ... '
	60	PRINT : PRINT Q$; : INPUT XD,XM,XS
65 . . . and convert to decimal form with MINSEC	65	SW(1)=-1 : GOSUB 1000 : GP=FNM(X)
70 Which direction?	70	Q$="Equatorial to horizon (Y or N) .... "
75 Call YESNO to see	75	PRINT : GOSUB 960
	80	IF E=1 THEN SW=1 : GOTO 135
	85	Q$="Horizon to equatorial (Y or N) .... "
	90	GOSUB 960 : SW=-1
95 Jump to the end if neither direction	95	IF E=0 THEN GOTO 235
100 Convert to equatorial coordinates	100	Q$="Azimuth (D,M,S; N=0) .... "
105 Get values	105	PRINT : PRINT Q$; : INPUT XD,XM,XS
110 Call MINSEC to convert to decimal form	110	SW(1)=-1 : GOSUB 1000 : XA=FNM(X)
	115	Q$="Altitude (D,M,S) .... "
	120	PRINT Q$; : INPUT XD,XM,XS
	125	SW(1)=-1 : GOSUB 1000 : Y=FNM(X) : X=XA
	130	GOTO 165
135 Convert to horizon coordinates	135	Q$="Hour angle (H,M,S) .... "
140 Get values	140	PRINT : PRINT Q$; : INPUT XD,XM,XS
145 Call MINSEC to convert to decimal form	145	SW(1)=-1 : GOSUB 1000 : XA=FNM(X*15.0)
	150	Q$="Declination (D,M,S) .... "
	155	PRINT Q$; : INPUT XD,XM,XS
	160	SW(1)=-1 : GOSUB 1000 : Y=FNM(X) : X=XA
165 Call EQHOR and convert results to degrees	165	GOSUB 1500 : P=FND(P) : Q=FND(Q)
	170	IF SW=1 THEN GOTO 205
175 Print results . . .	175	X=P/15.0 : NC=9 : SW(1)=1 : GOSUB 1000
	180	Q$="Hour angle (H,M,S) .... "
185 . . . neatly (columns lining up)	185	PRINT Q$+" "+MID$(OP$,4,12)
	190	X=Q : GOSUB 1000
	195	Q$="Declination (D,M,S) .... "
	200	PRINT Q$+OP$: GOTO 235
	205	X=P : NC=9 : SW(1)=1 : GOSUB 1000
	210	Q$="Azimuth (D,M,S; N=0) .... "
	215	PRINT Q$+OP$
	220	X=Q : GOSUB 1000
	225	Q$="Altitude (D,M,S) .... "
	230	PRINT Q$+OP$

235	Another calculation?...	235	`Q$="Again (Y or N) ......................... "`
240	YESNO provides the answer	240	`PRINT : GOSUB 960`
		245	`IF E=0 THEN STOP`
250	If FL(4)=0 we have not yet done anything	250	`IF FL(4)=0 THEN GOTO 50`
		255	`Q$="Same place (Y or N) ..................... "`
260	Set FL(4)=0 for a new location	260	`GOSUB 960 : FL(4)=E : GOTO 50`

```
INCLUDE YESNO, MINSEC, EQHOR
```

Example

```
Equatorial and horizon coords.
-------------------------------


Geographical latitude (D,M,S; S neg.) ... ? -33,27,45.26

Equatorial to horizon (Y or N) ......... ? Y

Hour angle (H,M,S) ..................... ? 23,21,44.5
Declination (D,M,S) .................... ? 49,6,55.77
Azimuth (D,M,S; N=0) ...................   +  6 17 25.94
Altitude (D,M,S) .......................   +  6 59  0.84

Again (Y or N) ......................... ? Y
Same place (Y or N) .................... ? Y

Equatorial to horizon (Y or N) ......... ? N
Horizon to equatorial (Y or N) ......... ? Y

Azimuth (D,M,S; N=0) ................... ? 6,17,25.94
Altitude (D,M,S) ....................... ? 6,59,0.84
Hour angle (H,M,S) .....................   23 21 44.50
Declination (D,M,S) ....................   + 49  6 55.77

Again (Y or N) ......................... ? Y
Same place (Y or N) .................... ? N

Geographical latitude (D,M,S; S neg.) ... ? 51,45,23.82

Equatorial to horizon (Y or N) ......... ? Y

Hour angle (H,M,S) ..................... ? -8,21,44.2
Declination (D,M,S) .................... ? 0,0,1.1
Azimuth  (D,M,S; N=0) ..................   + 60 48  3.99
Altitude (D,M,S) .......................   - 21  1 51.72

Again (Y or N) ......................... ? Y
Same place (Y or N) .................... ? Y

Equatorial to horizon (Y or N) ......... ? N
Horizon to equatorial (Y or N) ......... ? Y

Azimuth (D,M,S; N=0) ................... ? 60,48,3.99
Altitude (D,M,S) ....................... ? -21,1,51.72
Hour angle (H,M,S) .....................   15 38 15.80
Declination (D,M,S) ....................   +  0  0  1.10

Again (Y or N) ......................... ? W
What ?
Again (Y or N) ......................... ? N
```

1600 HRANG

This routine converts the right ascension into the hour angle, and vice-versa, given the Greenwich sidereal time and the geographical longitude. (See also GENCON.)

The problem of converting between the hour angle and the right ascension of a celestial object at a given local sidereal time is so straightforward as to make it hardly seem worthwhile devoting a separate subroutine to it. Nevertheless, it crops up sufficiently often that one quickly becomes tired of writing the few lines of code needed every time into a new program, and so I have included the routing HRANG here. The equation is the same for conversion in either direction, there being no need for a switch to specify it. The right ascension or hour angle (hours) is input via the variable X and the corresponding hour angle or right ascension (hours) returned by P. The geographical longitude, GL, must be specified in degrees (W negative) and the Greenwich sidereal time, SG, in hours. The routine also returns the local sidereal time, LS, in hours.

Formulae
$P = [LS - X]_{24}$
$LS = [SG + (GL/15)]_{24}$
P = hour angle or right ascension
given
X = right ascension or hour angle
SG = Greenwich sidereal time
GL = geographical longitude
$[\ ]_{24}$ indicates reduction to the range 0–24 by the addition or subtraction of multiples of 24.

Details of HRANG
Called by GOSUB 1600.
Converts X, the hour angle or right ascension, into P, the corresponding right ascension or hour angle, at SG, the Greenwich sidereal time, all quantities in *hours*. The geographical longitude, GL, must also be specified in *degrees* (W negative, E positive). The local sidereal time in hours is returned by LS.

```
1597 REM
1598 REM       Subroutine HRANG
1599 REM
```

1600	Calculate local sidereal time in range 0–24

```
1600 A=SG+(GL/15.0) : GOSUB 1615 : LS=A
```

1605	Convert between right ascension and hour angle in range 0–24

```
1605 A=LS-X : GOSUB 1615 : P=A
```

1610	Return from HRANG

```
1610 RETURN
```

1615	Local routine to return A in range 0–24

```
1615 IF A>24 THEN A=A-24.0 : GOTO 1615
1620 IF A<0 THEN A=A+24.0 : GOTO 1620
1625 RETURN
```

		1	REM
		2	REM　　　　Handling program HHRANG
		3	REM
10	JULDAY needs this function	5	DIM FL(20),ER(20),SW(20)
		10	DEF FNI(W)=SGN(W)*INT(ABS(W))
		15	PRINT : PRINT : SP=0 : SH=0 : FL(3)=0
		20	PRINT "Right ascensions & hour angles"
		25	PRINT "---------------------------"
		30	PRINT : PRINT
35	SP is local switch signifying same place	35	IF SP=1 THEN GOTO 75
40	Get location details . . .	40	Q$="Daylight saving (H ahead of zone t)　"
45	. . . hours ahead of zone time (DS)	45	PRINT : PRINT Q$; : INPUT DS
		50	Q$="Time zone (hours; West negative) ... "
55	. . . time zone (TZ=0 for UT zone)	55	PRINT Q$; : INPUT TZ
		60	Q$="Geog. longitude (D,M,S; W neg.) .... "
65	. . . geographical longitude (GL) . . .	65	PRINT Q$; : INPUT XD,XM,XS
70	. . . in decimal degrees via MINSEC	70	SW(1)=-1 : GOSUB 1000 : GL=X
75	FL(3)=1 indicates same date as before	75	IF FL(3)=1 THEN PRINT : GOTO 90
80	Get date	80	Q$="Calendar date (D,M,Y) .... "
		85	PRINT : PRINT Q$; : INPUT DY,MN,YR
90	SH is local switch to indicate same time	90	IF SH=1 THEN GOTO 110
		95	Q$="Local civil time (H,M,S) .... "
100	Get the local civil time TM . . .	100	PRINT Q$; : INPUT XD,XM,XS
105	. . . in decimal hours via MINSEC	105	SW(1)=-1 : GOSUB 1000 : TM=X
110	Call TIME to convert TM to Greenwich sidereal time	110	SW(2)=1 : GOSUB 1300
115	ER(1) indicates impossible date (JULDAY)	115	IF ER(1)=1 THEN GOTO 235
120	Don't repeat yourself, otherwise . . .	120	IF SH=1 THEN GOTO 155
125	. . . display all the times calculated by TIME	125	Q$="Universal time (H,M,S) .... "
		130	PRINT Q$+UT$
		135	Q$="Greenwich sidereal time (H,M,S) .... "
		140	PRINT Q$+SG$
		145	Q$="Local sidereal time (H,M,S) .... "
		150	PRINT Q$+TL$
155	Which direction for conversion?	155	Q$="RA to hour angle (Y or N) .... "
160	Use YESNO to find out	160	PRINT : GOSUB 960 : SW=1
		165	IF E=1 THEN GOTO 195
		170	Q$="Hour angle to RA (Y or N) .... "
175	SW is local switch to indicate direction	175	GOSUB 960 : SW=-1
180	Jump to end if neither direction	180	IF E=0 THEN GOTO 235
185	Get the hour angle . . .	185	Q$="Hour angle (H,M,S) .... "
		190	GOTO 200
195	. . . or the right ascension . . .	195	Q$="Right ascension (H,M,S) .... "
200	. . . convert to hours with MINSEC . . .	200	PRINT : PRINT Q$; :　INPUT XD,XM,XS
		205	SW(1)=-1 : GOSUB 1000
210	. . . and call HRANG	210	GOSUB 1600 : X=P : SW(1)=1 : NC=9 : GOSUB 1000
215	Print results in H,M,S form . . .	215	Q$="Right ascension (H,M,S) .... "
		220	IF SW=-1 THEN GOTO 230
		225	Q$="Hour angle (H,M,S) .... "
230	. . . neatly	230	PRINT Q$+" "+MID$(OP$,4,12)

235	Another conversion?	235	`Q$="Again (Y or N) ..................... "`
240	Call YESNO to discover	240	`PRINT : GOSUB 960`
		245	`IF E=0 THEN STOP`
250	If FL(3)=0 we have not yet done anything	250	`Q$="Same place (Y or N) ............... "`
		255	`GOSUB 960 : SP=E`
260	SP=0 if we want a different place	260	`Q$="Same date (Y or N) ............... "`
		265	`GOSUB 960 : FL(3)=E`
270	FL(3)=0 if we want a new date	270	`Q$="Same time (Y or N) ................ "`
		275	`GOSUB 960 : SH=E : GOTO 35`
280	SH=0 for a new time		

```
INCLUDE YESNO, MINSEC, JULDAY, TIME, HRANG
```

Example

```
Right ascensions & hour angles
-------------------------------


Daylight saving (H ahead of zone t)  ? 2
Time zone (hours; West negative) ...  ? 0
Geog. longitude (D,M,S; W neg.) ....  ? 2,21,03.3

Calendar date (D,M,Y) ..............  ? 29,9,1992
Local civil time (H,M,S) ...........  ? 13,56,44.37
Universal time (H,M,S) .............    11 56 44.37
Greenwich sidereal time (H,M,S) ....    12 30 42.66
Local sidereal time (H,M,S) ........    12 40  6.88

RA to hour angle (Y or N) .........  ? Y

Right ascension (H,M,S) ............  ? 23,21,0
Hour angle (H,M,S) .................    13 19  6.88

Again (Y or N) ....................  ? Y
Same place (Y or N) ...............  ? Y
Same date (Y or N) ................  ? Y
Same time (Y or N) ................  ? Y


RA to hour angle (Y or N) .........  ? N
Hour angle to RA (Y or N) .........  ? Y

Hour angle (H,M,S) .................  ? 13,19,6.88
Right ascension (H,M,S) ............    23 21  0.00

Again (Y or N) ....................  ? N
```

Here is a handling program which combines the functions of EQHOR and HRANG to make conversions directly between right ascension/declination and azimuth/altitude. It takes account of your location and the date, and hence may be used anywhere in the world. ☞

	1	REM
	2	REM Handling program HALTAZ
	3	REM
	5	DIM FL(20),ER(20),SW(20)
10 FNI needed by JULDAY	10	DEF FNI(W)=SGN(W)*INT(ABS(W))
15 FNM to convert degrees to radians	15	DEF FNM(W)=1.745329252E-2*W
20 FND to convert radians to degrees	20	DEF FND(W)=5.729577951E1*W
25 FNS returns inverse sine	25	DEF FNS(W)=ATN(W/(SQR(1-W*W)+1E-20))
30 FNC returns inverse cosine	30	DEF FNC(W)=1.570796327-FNS(W)
	35	PRINT : PRINT : SH=0
	40	PRINT "Altitudes and azimuths"
	45	PRINT "---------------------" : PRINT
50 FL(4)=1 indicates same place as before	50	IF FL(4)=1 THEN GOTO 105
55 Get details of the location ...	55	Q$="Daylight saving (H ahead of zone t) "
60 ...hours ahead of zone time (DS)	60	PRINT : PRINT Q$; : INPUT DS
	65	Q$="Time zone (hours; West negative) ... "
70 ... time zone (TZ=0 for UT zone)	70	PRINT Q$; : INPUT TZ
	75	Q$="Geog. longitude (D,M,S; W neg.) .... "
80 ... geographical longitude (GL)...	80	PRINT Q$; : INPUT XD,XM,XS
85 ... in decimal degrees via MINSEC	85	SW(1)=-1 : GOSUB 1000 : GL=X
	90	Q$="Geog. latitude (D,M,S; S neg.) ..... "
95 ... geographical latitude (GP)...	95	PRINT Q$; : INPUT XD,XM,XS
100 ... in radians via MINSEC and FNM	100	SW(1)=-1 : GOSUB 1000 : GP=FNM(X)
105 FL(3)=1 indicates same date as before	105	IF FL(3)=1 THEN PRINT : GOTO 120
	110	Q$="Calendar date (D,M,Y) .... "
115 Get a new date	115	PRINT : PRINT Q$; : INPUT DY,MN,YR
120 SH is a local switch indicating same time if set to 1	120	IF SH=1 THEN GOTO 140
	125	Q$="Local civil time (H,M,S) .... "
130 Get a new local civil time (TM)...	130	PRINT Q$; : INPUT XD,XM,XS
135 ... in decimal hours via MINSEC	135	SW(1)=-1 : GOSUB 1000 : TM=X
140 Call TIME to convert to sidereal	140	SW(2)=1 : GOSUB 1300
145 ER(1)=1 if date is impossible	145	IF ER(1)=1 THEN GOTO 375
150 Don't repeat ourselves, otherwise...	150	IF SH=1 THEN GOTO 185
155 ... display times calculated by TIME	155	Q$="Universal time (H,M,S) .... "
	160	PRINT Q$+UT$
	165	Q$="Greenwich sidereal time (H,M,S) .... "
	170	PRINT Q$+SG$
	175	Q$="Local sidereal time (H,M,S) .... "
	180	PRINT Q$+TL$
185 Which direction to convert?...	185	Q$="Equatorial to horizon (Y or N) ..... "
190 Call YESNO for the answer	190	PRINT : GOSUB 960
	195	IF E=1 THEN GOTO 295
	200	Q$="Horizon to equatorial (Y or N) ..... "
	205	GOSUB 960
210 Jump to the end if neither direction	210	IF E=0 THEN GOTO 375
215 ... to equatorial coordinates	215	Q$="Azimuth (D,M,S; N=0) .... "
220 Get the coordinates...	220	PRINT : PRINT Q$; : INPUT XD,XM,XS
225 ... in radians	225	SW(1)=-1 : GOSUB 1000 : XA=FNM(X)
	230	Q$="Altitude (D,M,S) .... "
	235	PRINT Q$; : INPUT XD,XM,XS
	240	SW(1)=-1 : GOSUB 1000 : Y=FNM(X) : X=XA

245	Call EQHOR to convert to hour angle and declination	245
250	Display hour angle in H,M,S form via MINSEC...	250
260	...neatly	255
265	Call HRANG to convert to right ascension...	260
		265
270	...and display in H,M,S form via MINSEC...	270
275	...neatly	275
280	Display declination in D,M,S form	280
		285
290	Skip to the end	290

```
245   GOSUB 1500 : P=FND(P) : Q=FND(Q)

250   X=P/15.0 : NC=9 : SW(1)=1 : GOSUB 1000

255   Q$="Hour angle (H,M,S) ................. "
260   PRINT Q$+" "+MID$(OP$,4,12)
265   GOSUB 1600 : X=P : GOSUB 1000

270   Q$="Right ascension (H,M,S) ........... "

275   PRINT Q$+" "+MID$(OP$,4,12)
280   X=Q : GOSUB 1000
285   Q$="Declination (D,M,S) ............... "
290   PRINT Q$+OP$ : GOTO 375
```

295	...to horizon coordinates	295
300	Get the right ascension...	300
305	...convert to decimal hours (MINSEC) and call HRANG...	305
310	...convert hour angle to H,M,S form with MINSEC...	310
315	...and display it...	315
320	...neatly	320
325	Get the declination...	325
		330
335	...in radians; convert hour angle to radians	335

```
295   Q$="Right ascension (H,M,S) ........... "
300   PRINT : PRINT Q$; : INPUT XD,XM,XS
305   SW(1)=-1 : GOSUB 1000 : GOSUB 1600

310   X=P : NC=9 : SW(1)=1 : GOSUB 1000

315   Q$="Hour angle (H,M,S) ................. "
320   PRINT Q$+" "+MID$(OP$,4,12)
325   Q$="Declination (D,M,S) ............... "
330   PRINT Q$; : INPUT XD,XM,XS
335   SW(1)=-1 : GOSUB 1000 : Y=FNM(X) : X=FNM(P*15.0)
```

340	Call EQHOR to find azimuth and altitude	340
345	Convert azimuth to D,M,S form...	345
		350
355	...and display it	355
360	Convert altitude to D,M,S form...	360
		365
370	...and display it	370

```
340   GOSUB 1500 : P=FND(P) : Q=FND(Q)

345   X=P : NC=9 : SW(1)=1 : GOSUB 1000
350   Q$="Azimuth   (D,M,S; N=0) ............ "
355   PRINT Q$+OP$
360   X=Q : GOSUB 1000
365   Q$="Altitude (D,M,S) .................. "
370   PRINT Q$+OP$
```

375	Another conversion?	375
380	Call YESNO for a yes or a no	380
		385
390	If we have not yet converted anything, don't set switches	390
		395
400	FL(4)=0 for a new place	400
		405
410	FL(3)=0 for a new date	410
		415
420	SH=0 for a new time	420

```
375   Q$="Again (Y or N) .................... "
380   PRINT : GOSUB 960
385   IF E=0 THEN STOP
390   IF FL(4)=0 THEN GOTO 50

395   Q$="Same place (Y or N) ............... "
400   GOSUB 960 : FL(4)=E
405   Q$="Same date (Y or N) ................ "
410   GOSUB 960 : FL(3)=E
415   Q$="Same time (Y or N) ................ "
420   GOSUB 960 : SH=E : GOTO 50
```

```
INCLUDE YESNO, MINSEC, JULDAY, TIME, EQHOR, HRANG
```

1600 HRANG

Example

```
Altitudes and azimuths
----------------------


Daylight saving (H ahead of zone t)  ? 0
Time zone (hours; West negative) ...  ? -5
Geog. longitude (D,M,S; W neg.) ....  ? -77,0,0
Geog. latitude (D,M,S; S neg.) .....  ? 38,55,0

Calendar date (D,M,Y) ..............  ? 1,2,1984
Local civil time (H,M,S) ...........  ? 7,23,0
Universal time (H,M,S) .............    12 23  0.00
Greenwich sidereal time (H,M,S) ....    21  6 37.98
Local sidereal time (H,M,S) ........    15 58 37.98

Equatorial to horizon (Y or N) .....  ? Y

Right ascension (H,M,S) ............  ? 20,40,5.2
Hour angle (H,M,S) .................    19 18 32.78
Declination (D,M,S) ................  ? -22,12,0
Azimuth  (D,M,S; N=0) ..............    +119 18 14.59
Altitude (D,M,S) ...................    +  0 16 15.88

Again (Y or N) .....................  ? Y
Same place (Y or N) ................  ? Y
Same date (Y or N) .................  ? Y
Same time (Y or N) .................  ? Y


Equatorial to horizon (Y or N) .....  ? N
Horizon to equatorial (Y or N) .....  ? Y

Azimuth (D,M,S; N=0) ...............  ? 119,18,14.59
Altitude (D,M,S) ...................  ? 0,16,15.88
Hour angle (H,M,S) .................    19 18 32.78
Right ascension (H,M,S) ............    20 40  5.20
Declination (D,M,S) ................    - 22 12  0.00

Again (Y or N) .....................  ? N
```

1700 OBLIQ

This routine calculates the value of the obliquity of the ecliptic for a given calendar date.

As the Earth moves around the Sun, it sweeps out the *plane of the ecliptic*, making one complete circuit in one year. The Earth also rotates about its own North–South polar axis, making one revolution in 24 hours. Since the Earth acts like a huge gyroscope, the direction of its polar axis remains relatively-fixed, although *precession* (PRCESS1, 2500; PRCESS2, 2600) and *nutation* (NUTAT, 1800) conspire to cause a wobbling motion. This direction is inclined at about 23.5 degrees to the perpendicular to the plane of the ecliptic, and hence the Earth's equator is tilted at the same angle to the plane of the ecliptic. The angle is called the *obliquity of the ecliptic* and it is needed whenever we convert between ecliptic and equatorial coordinates (see EQECL, 2000). The angle changes slowly with time so it must be determined separately for calculations made at different epochs.

The subroutine OBLIQ returns the value of the obliquity of the ecliptic via the variable OB (degrees). The date is input in the usual way via DY (days), MN (months), and YR (years). Execution is controlled by the flag FL(5), designed to save time if the calculation has already been made for the current date. When calling the routine for the first time, or for a new date, set FL(5) = 0. It is set to 1 by the routine (line 1705). Subsequent calls to OBLIQ with FL(5) = 1 result in no calculation being performed, but control is returned immediately to the calling program. Note that OBLIQ calls NUTAT (line 1705) to include nutation. If you do not wish to include nutation, you must set FL(6)=1 and DO=0 before calling OBLIQ. Otherwise, a new value of DO is calculated and added to OB (line 1725). If the date has changed since the last call, you must remember to reset FL(6)=0 for a new value of DO.

1700 OBLIQ

Formulae
$OB = 23.43929167 - (A/3600) + DO$
$A = (4.6185E1 \times T) + (6.0E-4 \times T^2) - (1.81E-3 \times T^3)$
$T = (DJ/36525) - 1$
DJ = Julian days since 1900 January 0.5
DO = nutation in obliquity (NUTAT)

Details of OBLIQ
Called by GOSUP 1700.

Calculates OB (degrees), the obliquity of the ecliptic for the date given by DY (days), MN (months), and YR (years).

Set FL(5) = 0 on first call and for new date. FL(5) is set to 1 by the routine. If FL(5) = 1 on call, then execution is returned immediately to the calling program without calculation.

Set FL(6) = 1 if nutation must not be included.

Other routines called: JULDAY (1100), NUTAT (1800).

```
1697   REM
1698   REM        Subroutine OBLIQ
1699   REM
```

1700	Obliquity already calculated for this date?

```
1700   IF FL(5)=1 THEN RETURN
```

1705	Call NUTAT to find DO and JULDAY to find DJ
1710	Impossible date?
1715	T is Julian centuries since 2000 January 1.5

```
1705   FL(5)=1 : FL(1)=0 : GOSUB 1800 : GOSUB 1100
1710   IF ER(1)=1 THEN FL(5)=0 : RETURN
1715   T=(DJ/36525.0)-1.0

1720   A=(46.815+(0.0006-0.00181*T)*T)*T
```

| 1725 | Find obliquity, OB, in degrees |
| 1730 | Normal return from OBLIQ |

```
1725   A=A/3600.0 : OB=23.43929167-A+DO
1730   RETURN
```

```
1      REM
2      REM        Handling program HOBLIQ
3      REM

5      DIM FL(20),ER(20),SW(20)
```

| 10 | FNI needed by JULDAY |
| 15 | Degrees to radians conversion |

```
10     DEF FNI(W)=SGN(W)*INT(ABS(W))
15     DEF FNM(W)=1.745329252E-2*W

20     PRINT : PRINT
25     PRINT "Obliquity of the ecliptic"
30     PRINT "-------------------------"
35     PRINT : FL(1)=0 : FL(5)=0 : FL(6)=0
```

| 40 | Get date |

```
40     Q$="Calendar date (D,M,Y) ..... "
45     PRINT Q$; : INPUT DY,MN,YR
```

50	Include nutation in obliquity as well?
55	Call YESNO to find out
60	Set FL(6)=1 for no nutation

```
50     Q$="Include nutation (Y/N) .... "
55     GOSUB 960
60     IF E=0 THEN FL(6)=1 : DO=0.0
```

| 65 | Call OBLIQ |
| 70 | ER(1)=1 if date is impossible (see JULDAY) |

```
65     GOSUB 1700
70     IF ER(1)=1 THEN GOTO 90
```

| 75 | Convert obliquity, OB, to D,M,S form with MINSEC |

```
75     X=OB : NC=9 : SW(1)=1 : GOSUB 1000
```

| 85 | Print the result |

```
80     Q$="The obliquity is (D,M,S) .. "
85     PRINT Q$+OP$
```

| 90 | Another date? |
| 95 | YESNO tells us |

```
90     Q$="Again (Y or N) ........... "
95     PRINT : GOSUB 960
100    IF E=0 THEN STOP
105    GOTO 35

INCLUDE YESNO, MINSEC, JULDAY, OBLIQ, NUTAT
```

1700 OBLIQ

Example

```
Obliquity of the ecliptic
-------------------------

Calendar date (D,M,Y) ..... ? 4,2,1989
Include nutation (Y/N) .... ? N
The obliquity is (D,M,S) ..   + 23 26 26.56

Again (Y or N) ........... ? Y

Calendar date (D,M,Y) ..... ? 4,2,1989
Include nutation (Y/N) .... ? Y
The obliquity is (D,M,S) ..   + 23 26 34.79

Again (Y or N) ........... ? Y

Calendar date (D,M,Y) ..... ? 1.5,1,2000
Include nutation (Y/N) .... ? N
The obliquity is (D,M,S) ..   + 23 26 21.45

Again (Y or N) ........... ? Y

Calendar date (D,M,Y) ..... ? 1.5,1,2000
Include nutation (Y/N) .... ? Y
The obliquity is (D,M,S) ..   + 23 26 15.56

Again (Y or N) ........... ? N
```

The Astronomical Almanac lists the value of the obliquity of the ecliptic for 2000 January 1.5 as 23° 26′ 21″.448.

1800 NUTAT

This routine calculates the nutation in ecliptic longitude and nutation in the obliquity of the ecliptic for a given date.

The combined gravitational fields of the Sun and the Moon acting on the non-spherical Earth cause the direction of the Earth's rotation axis to gyrate slowly with a period of about 25 800 years. This effect, called *precession*, is calculated by routines PRCESS1 (2500), and PRCESS2 (2600). Superimposed on the regular motion there are also small additional periodic terms caused by the varying distances and relative directions of the Moon and Sun, which continuously alter the strength and direction of the gravitational field. This slight wobbling motion is called *nutation*, and it must be taken into account when accuracies of better than half an arcminute are required. The routine given here calculates the effects of nutation on the ecliptic longitude (DP) and on the obliquity of the ecliptic (DO). It has an intrinsic accuracy (reduced by lack of precision in your machine) of about 1 arcsecond.

The calculations are made by the routine for the date input as DY (days), MN (months), and YR (years). Execution is controlled by the flag FL(6). If this flag is set to 0 before calling NUTAT, execution of the routine continues normally. It is set to 1 at line 1865. Subsequent calls to NUTAT with FL(6) = 1 do not result in any new calculations, but control is returned immediately to the calling program. This is to avoid recalculating the nutation for the same date when several routines are strung together in a single program, each of which makes a call to NUTAT. FL(6) should be set to zero when calling the routine for the first time, and whenever the nutation for a new date is required. FL(6) should be preset to 1, DP to 0, and DO to 0 if nutation should not be taken into account.

Formulae

DP and DO are calculated in terms of the Sun's mean longitude, L1, the Moon's mean longitude, D1, the Sun's mean anomaly, M1, the Moon's mean anomaly, M2, and the longitude of the Moon's ascending node, N1. (See the listing of the routine for the various coefficients and arguments of the trigonometric functions.) Each is dependent on T, the number of Julian centuries of 36 525 days since 1900 January 0.5. For example, the Sun's mean longitude is given in degrees by

$$L1 = 279.6967 + (36000.7689 \times T) + (0.000303 \times T^2).$$

This is a large number for common values of T, yet must be reduced accurately near to the range 0–360 degrees by the addition or subtraction of multiples of 360. Better accuracy is obtained if the reduction is done before multiplying by T. Thus it may be expressed as

$$L1 = 2.796967E2 + (3.03E{-}4 \times T^2) + B,$$
$$B = 360 \times (A - INT(A)),$$
$$A = 1.000021358E2 \times T,$$

where INT may represent either the least-integer or truncated-integer function (with slightly different but equally-valid results for negative T, one within the range 0–360 and the other just outside it). This is the form used by NUTAT.

Details of NUTAT

Called by GOSUB 1800.

Calculates DO,† the nutation in obliquity, and DP, the nutation in longitude, both in degrees, for the date given by DY (days), MN (months), and YR (years).

Set FL(6) to 0 on first call and for new date. FL(6) is set to 1 by the routine. If FL(6) = 1 on call, control is returned immediately to the calling program without a new calculation of DO and DP.

Other routine called: JULDAY (1100).

---

† (note added in proof). In some compilers and interpreters, DO is a reserved word. Substitute, say, DDO for DO here and elsewhere throughout the book if yours gives trouble.

```
1797 REM
1798 REM        Subroutine NUTAT
1799 REM
```

1800 FL(6) set to 0 for a new calculation

```
1800 IF FL(6)=1 THEN RETURN
```

1805 Call JULDAY for DJ; T in centuries since epoch

```
1805 FL(1)=0 : GOSUB 1100 : T=DJ/36525.0 : T2=T*T
```

1810 ER(1)=1 if impossible date

```
1810 IF ER(1)=1 THEN FL(6)=0 : RETURN
1815 A=1.000021358E2*T : B=360.0*(A-INT(A))
```

1820 L1 is the Sun's mean longitude

```
1820 L1=2.796967E2+3.03E-4*T2+B : L2=2.0*FNM(L1)
1825 A=1.336855231E3*T : B=360.0*(A-INT(A))
```

1830 D1 is the Moon's mean longitude

```
1830 D1=2.704342E2-1.133E-3*T2+B : D2=2.0*FNM(D1)
1835 A=9.999736056E1*T : B=360.0*(A-INT(A))
```

1840 M1 is the Sun's mean anomaly

```
1840 M1=3.584758E2-1.5E-4*T2+B : M1=FNM(M1)
1845 A=1.325552359E3*T : B=360.0*(A-INT(A))
```

1850 M2 is the Moon's mean anomaly

```
1850 M2=2.961046E2+9.192E-3*T2+B : M2=FNM(M2)
1855 A=5.372616667*T : B=360.0*(A-INT(A))
```

1860 N1 is the longitude of the Moon's ascending node

```
1860 N1=2.591833E2+2.078E-3*T2-B : N1=FNM(N1)
```

1865 ... all in radians

```
1865 N2=2*N1 : FL(6)=1
```

1870 Calculate the nutation in longitude ...

```
1870 DP=(-17.2327-1.737E-2*T)*SIN(N1)
1875 DP=DP+(-1.2729-1.3E-4*T)*SIN(L2)+2.088E-1*SIN(N2)
1880 DP=DP-2.037E-1*SIN(D2)+(1.261E-1-3.1E-4*T)*SIN(M1)
1885 DP=DP+6.75E-2*SIN(M2)-(4.97E-2-1.2E-4*T)*SIN(L2+M1)
1890 DP=DP-3.42E-2*SIN(D2-N1)-2.61E-2*SIN(D2+M2)
1895 DP=DP+2.14E-2*SIN(L2-M1)-1.49E-2*SIN(L2-D2+M2)
1900 DP=DP+1.24E-2*SIN(L2-N1)+1.14E-2*SIN(D2-M2)
```

1905 Calculate the nutation in obliquity ...

```
1905 DO=(9.21+9.1E-4*T)*COS(N1)
1910 DO=DO+(5.522E-1-2.9E-4*T)*COS(L2)-9.04E-2*COS(N2)
1915 DO=DO+8.84E-2*COS(D2)+2.16E-2*COS(L2+M1)
1920 DO=DO+1.83E-2*COS(D2-N1)+1.13E-2*COS(D2+M2)
1925 DO=DO-9.3E-3*COS(L2-M1)-6.6E-3*COS(L2-N1)
```

1930 ... both in degrees, and return from NUTAT

```
1930 DP=DP/3600.0 : DO=DO/3600.0 : RETURN
```

```
1      REM
2      REM          Handling program HNUTAT
3      REM

5      DIM FL(20),ER(20),SW(20)
10     DEF FNI(W)=SGN(W)*INT(ABS(W))
15     DEF FNM(W)=1.745329252E-2*W

20     PRINT : PRINT
25     PRINT "Nutation in obliquity and longitude"
30     PRINT "-----------------------------------"
35     PRINT : FL(1)=0 : FL(6)=0

40     Q$="Calendar date (D,M,Y) ....... "
45     PRINT Q$; : INPUT DY,MN,YR

50     GOSUB 1800
55     IF ER(1)=1 THEN GOTO 85

60     X=DP : NC=9 : SW(1)=1 : GOSUB 1000
65     Q$="Nutation in longitude (D,M,S) "

70     PRINT Q$+OP$ : X=DO : GOSUB 1000
75     Q$="Nutation in obliquity (D,M,S) "
80     PRINT Q$+OP$

85     Q$="Again (Y or N) ............. "
90     PRINT : GOSUB 960
95     IF E=0 THEN STOP
100    GOTO 35
```

The notes alongside the listing read:

- **10** FNI is needed by JULDAY
- **15** Degrees to radians conversion
- **40** Get the date . . .
- **50** . . . and call NUTAT
- **55** Check for impossible date
- **60** Convert results to D,M,S form using MINSEC . . .
- **70** . . . and display
- **85** Another date?
- **90** YESNO gets the answer

INCLUDE YESNO, MINSEC, JULDAY, NUTAT

Example

```
Nutation in obliquity and longitude
-----------------------------------

Calendar date (D,M,Y) ....... ? 4,2,1989
Nutation in longitude (D,M,S)   +  0  0  8.30
Nutation in obliquity (D,M,S)   +  0  0  8.26

Again (Y or N) ............. ? y

Calendar date (D,M,Y) ....... ? 1.5,1,2000
Nutation in longitude (D,M,S)   -  0  0 13.96
Nutation in obliquity (D,M,S)   -  0  0  5.76

Again (Y or N) ............. ? y

Calendar date (D,M,Y) ....... ? 23,4,1995
Nutation in longitude (D,M,S)   +  0  0  9.53
Nutation in obliquity (D,M,S)   -  0  0  7.25

Again (Y or N) ............. ? n
```

58

2000 EQECL

This routine converts geocentric ecliptic longitude and latitude into the corresponding right ascension and declination, and vice-versa. (See also GENCON.)

The positions of members of our Solar System may best be specified in the ecliptic coordinate system (see Figure 4) since this uses the plane of the ecliptic as the fundamental plane and most members of the Solar System move in orbits close to it. The reference direction is that of the vernal equinox, as in the equatorial system. This is an obvious choice because it lies along the line of intersection of the planes of the Earth's equator and the ecliptic. The ecliptic longitude is the angle measured in the ecliptic round from the vernal equinox in the same sense as that in which the right ascension is measured, and the ecliptic latitude is the angle up from the ecliptic, positive if to the north, or 'above' the ecliptic, and negative if to the south, or 'below' the ecliptic.

The subroutine EQECL converts coordinates between the ecliptic and equatorial systems. The angles to be converted are input via the variables X (ecliptic longitude or right ascension) and Y (ecliptic latitude or declination), with the results output via P (right ascension or ecliptic longitude) and Q (declination or ecliptic latitude). The formulae are not quite symmetrical in the two pairs of coordinates so that the direction of conversion needs to be specified with a switch. Set SW(3) to +1 for conversion from equatorial to ecliptic

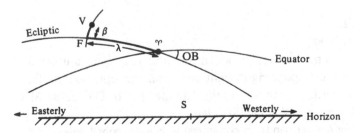

Figure 4. Ecliptic coordinates as seen from the ground. S is the South point of the observer's horizon. The planes of the ecliptic and equator intersect at ♈, the vernal equinox. VF is an arc of the great circle through the north pole of the ecliptic and the planet V, intersecting the equator at F. λ and β are the ecliptic longitude and latitude of V respectively.

coordinates, and to −1 for conversion from ecliptic to equatorial coordinates. The routine also calls OBLIQ (1700) to find the mean value of the obliquity of the ecliptic, OB, and OBLIQ itself calls NUTAT to calculate nutation in ecliptic longitude and obliquity. The latter is added automatically to OB. If you do not wish to include nutation in obliquity, therefore, you should preset FL(6) to 1 and set DO to 0 before calling EQECL. The date should be input in the usual way via DY (days), MN (months) and YR (years).

Execution of EQECL is controlled by the flag FL(7). This must be set to 0 when calling the subroutine for the first time, or whenever a new date is specified. FL(7) is set to 1 by the routine and subsequent calls with FL(7) = 1 by-pass the calculations in lines 2005–2020 inclusive which depend only on the date. Hence execution time is saved.

Formulae

$$TAN(P) = ((SIN(X) \times COS(OB)) + (TAN(Y) \times SIN(OB) \times SW(3)))/ \\ COS(X)$$

$$SIN(Q) = (SIN(Y) \times COS(OB)) - (COS(Y) \times SIN(OB) \times SIN(X) \\ \times SW(3))$$

OB = obliquity of the ecliptic

If SW(3) = +1:

P = ecliptic longitude

Q = ecliptic latitude

X = right ascension

Y = declination

If SW(3) = −1:

P = right ascension

Q = declination

X = ecliptic longitude

Y = ecliptic latitude

Details of EQECL

Called by GOSUB 2000.

Converts X, ecliptic longitude or right ascension, and Y, ecliptic latitude or declination, into the corresponding P, right ascension or ecliptic longitude, and Q, declination or ecliptic latitude, for the date specified by DY (days), MN (months), and YR (years). All angles in *radians*.

Set SW(3) = +1 for conversion from equatorial to ecliptic coordinates.

Set SW(3) = −1 for conversion from ecliptic to equatorial coordinates.

Set FL(7) = 0 on first call or for conversion on a new date. FL(7) is set to 1 by the routine. Subsequent calls with FL(7) = 1 by-pass calculations involving only the date and hence save time.

Set FL(6) to 1 and DO to 0 if nutation is not to be included.

Other routine called: OBLIQ (1700).

```
                        1997  REM
                        1998  REM        Subroutine EQECL
                        1999  REM
```

2000 First call for this date?
```
                        2000  IF FL(7)=1 THEN GOTO 2025
```

2005 Set FL(7)=1 to by-pass this bit (reset for new date)
```
                        2005  PI=3.1415926535 : TP=2*PI : FL(7)=1
```

2010 Call OBLIQ for new value of the obliquity OB
```
                        2010  FL(5)=0 : GOSUB 1700
```

2015 Impossible date? If so, abort
```
                        2015  IF ER(1)=1 THEN FL(7)=0 : RETURN
```

2020 Find sine and cosine of OB (radians)
```
                        2020  E=FNM(OB) : SE=SIN(E) : CE=COS(E)
```

2025 Find trigonometric values once only
```
                        2025  CY=COS(Y) : SY=SIN(Y)
```

2030 CY always positive; make sure it's not zero
```
                        2030  IF ABS(CY)<1E-20 THEN CY=1E-20
                        2035  TY=SY/CY : CX=COS(X) : SX=SIN(X)
```

2040 SW(3) signifies direction of conversion
```
                        2040  S=(SY*CE)-(CY*SE*SX*SW(3))
```

2045 FNS is inverse sine function
```
                        2045  Q=FNS(S) : A=(SX*CE)+(TY*SE*SW(3))
                        2050  P=ATN(A/CX)
```

2055 Remove ambiguity of inverse tangent . . .
```
                        2055  IF CX<0 THEN P=P+PI
```

2060 . . . and reduce P to range 0–π (radians)
```
                        2060  IF P>TP THEN P=P-TP : GOTO 2060
                        2065  IF P<0 THEN P=P+TP : GOTO 2065
```

2070 Normal return from EQECL
```
                        2070  RETURN
```

		1	REM
		2	REM Handling program HEQECL
		3	REM
		5	DIM FL(20),SW(20),ER(20)
10	FNI is truncated-integer function needed by JULDAY	10	DEF FNI(W)=SGN(W)*INT(ABS(W))
15	Degrees to radians converter	15	DEF FNM(W)=1.745329252E-2*W
20	Radians to degrees converter	20	DEF FND(W)=5.729577951E1*W
25	FNS returns the inverse sine	25	DEF FNS(W)=ATN(W/(SQR(1-W*W)+1E-20))
		30	PRINT : PRINT : FL(7)=0
		35	PRINT "Equatorial and ecliptic coords."
		40	PRINT "-----------------------------"
		45	PRINT : PRINT
50	FL(7)=0 for a new date	50	IF FL(7)=1 THEN GOTO 80
55	Get a new date	55	Q$="Calendar date (D,M,Y) .... "
		60	PRINT : PRINT Q$; : INPUT DY,MN,YR
65	Do we take account of nutation?	65	Q$="Include nutation in obl. (Y or N) .. "
70	Call YESNO for the answer	70	GOSUB 960 : FL(6)=0
75	This what to do if you don't want nutation	75	IF E=0 THEN FL(6)=1 : DO=0
80	Which direction?	80	Q$="Equatorial to ecliptic (Y or N) .... "
85	YESNO tells us	85	PRINT : GOSUB 960
		90	IF E=1 THEN GOTO 180
		95	Q$="Ecliptic to equatorial (Y or N) .... "
		100	GOSUB 960
105	Skip to the end if neither	105	IF E=0 THEN GOTO 250
110	Ecliptic to equatorial	110	Q$="Ecliptic longitude (D,M,S) .... "
115	Get the longitude . . .	115	PRINT : PRINT Q$; : INPUT XD,XM,XS
120	. . . in radians via MINSEC and FNM	120	SW(1)=-1 : GOSUB 1000 : XA=FNM(X)
125	Get the latitude . . .	125	Q$="Ecliptic latitude (D,M,S) .... "
		130	PRINT Q$; : INPUT XD,XM,XS
135	. . . also in radians	135	SW(1)=-1 : GOSUB 1000 : Y=FNM(X) : X=XA
140	Call EQECL, and convert results to degrees	140	SW(3)=-1 : GOSUB 2000 : P=FND(P) : Q=FND(Q)
145	Impossible date flag	145	IF ER(1)=1 THEN GOTO 250
150	Display the results in H,M,S form via MINSEC	150	X=P/15.0 : NC=9 : SW(1)=1 : GOSUB 1000
		155	Q$="Right ascension (H,M,S) .... "
160	. . . neatly	160	PRINT Q$+" "+MID$(OP$,4,12)
		165	X=Q : GOSUB 1000
		170	Q$="Declination (D,M,S) .... "
175	Display declination, and skip to the end	175	PRINT Q$+OP$: GOTO 250
180	Equatorial to ecliptic	180	Q$="Right ascension (H,M,S) .... "
185	Get the right ascension . . .	185	PRINT : PRINT Q$; : INPUT XD,XM,XS
190	. . . in radians via MINSEC and FNM	190	SW(1)=-1 : GOSUB 1000 : XA=FNM(X*15.0)
195	Get the declination . . .	195	Q$="Declination (D,M,S) .... "
		200	PRINT Q$; : INPUT XD,XM,XS
205	. . . in radians	205	SW(1)=-1 : GOSUB 1000 : Y=FNM(X) : X=XA
210	Call EQECL, and convert results to degrees	210	SW(3)=1 : GOSUB 2000 : P=FND(P) : Q=FND(Q)
215	Impossible date?	215	IF ER(1)=1 THEN GOTO 250
220	Convert results to H,M,S form . . .	220	X=P : NC=9 : SW(1)=1 : GOSUB 1000
		225	Q$="Ecliptic longitude (D,M,S) .... "

2000 EQECL

```
230   PRINT Q$+OP$
235   X=Q : GOSUB 1000
240   Q$="Ecliptic latitude (D,M,S) ......... "
245   PRINT Q$+OP$

250   Q$="Again (Y or N) ..................... "
255   PRINT : GOSUB 960
260   IF E=0 THEN STOP
265   IF FL(7)=0 THEN GOTO 50
270   Q$="Same date (Y or N) ................ "
275   GOSUB 960 : FL(7)=E : GOTO 50
```

INCLUDE YESNO, MINSEC, JULDAY, OBLIQ, NUTAT, EQECL

Example

```
Equatorial and ecliptic coords.
-------------------------------

Calendar date (D,M,Y) .............. ? 19,5,1950
Include nutation in obl. (Y or N) .. ? N

Equatorial to ecliptic (Y or N) .... ? Y

Right ascension (H,M,S) ............ ? 14,26,57
Declination (D,M,S) ................ ? 32,21,05
Ecliptic longitude (D,M,S) .........   +200 19  6.66
Ecliptic latitude (D,M,S) ..........   + 43 47 13.83

Again (Y or N) ..................... ? Y
Same date (Y or N) ................. ? Y

Equatorial to ecliptic (Y or N) .... ? N
Ecliptic to equatorial (Y or N) .... ? Y

Ecliptic longitude (D,M,S) ......... ? 200,19,6.66
Ecliptic latitude (D,M,S) .......... ? 43,47,13.83
Right ascension (H,M,S) ............   14 26 57.00
Declination (D,M,S) ................   + 32 21  5.00

Again (Y or N) ..................... ? Y
Same date (Y or N) ................. ? N

Calendar date (D,M,Y) .............. ? 28,5,2004
Include nutation in obl. (Y or N) .. ? Y

Equatorial to ecliptic (Y or N) .... ? Y

Right ascension (H,M,S) ............ ? 0,0,5.5
Declination (D,M,S) ................ ? -87,12,12
Ecliptic longitude (D,M,S) .........   +277  0  4.40
Ecliptic latitude (D,M,S) ..........   - 66 24 13.10

Again (Y or N) ..................... ? Y
Same date (Y or N) ................. ? Y

Equatorial to ecliptic (Y or N) .... ? N
Ecliptic to equatorial (Y or N) .... ? Y

Ecliptic longitude (D,M,S) ......... ? 277,0,4.4
Ecliptic latitude (D,M,S) .......... ? -66,24,13.10
Right ascension (H,M,S) ............    0  0  5.50
Declination (D,M,S) ................   - 87 12 12.00

Again (Y or N) ..................... ? N
```

2100 EQGAL

This routine converts galactic longitude and latitude into right ascension and declination, and vice-versa. (See also GENCON.)

The positions of objects in the Galaxy such as stars, globular clusters, pulsars etc. may best be described using the galactic coordinate system (see Figure 5). This system takes the plane of the Galaxy as its fundamental plane and the line joining the Sun to the centre of the Galaxy as the reference direction. The galactic longitude is the angle measured *at the Sun* in the galactic plane round from the direction of the galactic centre in the same sense as increasing right ascension. The galactic latitude is the angle up from the galactic plane, positive towards the north and negative towards the south.

Routine EQGAL converts between galactic and equatorial coordinates in either direction. The galactic longitude or right ascension is input via the variable X and the galactic latitude or declination via Y. The results are returned via P, the corresponding right ascension or galactic longitude, and Q, the declination or galactic latitude. The equations are not symmetrical in the two pairs

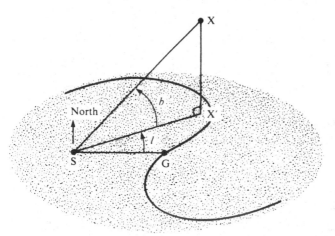

Figure 5. Galactic coordinates. The galactic longitude is the angle, *l*, in the plane of the Galaxy round from the line joining the Sun, S, to the galactic centre, G. The galactic latitude, *b*, is the angle from the plane.

2100 EQGAL

of coordinates so that the direction of conversion must be specified with a switch. SW(4) is given the value +1 for equatorial to galactic conversion, and −1 for galactic to equatorial conversion. The equations assume coordinates for the north galactic pole of right ascension 192.25 degrees and declination 27.4 degrees, and that the galactic longitude of the ascending node of the galactic plane on the equator is 33 degrees.

Execution of EQGAL is controlled by the flag FL(8). This must be set to 0 on the first call to the routine, but thereafter can be ignored. The flag is set to 1 in line 2115. Subsequent calls skip lines 2105–2115 inclusive which depend only on the data given at the end of the previous paragraph.

Formulae

$\text{TAN}(P) = (C/D) + A$

$\text{SIN}(Q) = (\text{COS}(Y) \times \text{COS}(GD) \times B) + (\text{SIN}(Y) \times \text{SIN}(GD))$

If SW(4) = +1:

  P = galactic longitude

  Q = galactic latitude

  X = right ascension

  Y = declination

  A = X − GR

  B = COS(A)

  C = SIN(Y) − (SIN(Q) × SIN(GD))

  D = COS(Y) × SIN(A) × GOS(GD)

  A = AN

If SW(4) = −1:

  P = right ascension

  Q = declination

  X = galactic longitude

  Y = galactic latitude

  A = X − AN

  B = SIN(A)

  C = COS(Y) × COS(A)

  D = (SIN(Y) × COS(GD)) − (COS(Y) × SIN(GD) × SIN(A))

  A = GR

In both cases:

  GR = 192.25 degrees

  GD = 27.4 degrees

  AN = 33.0 degrees

Details of EQGAL

Called by GOSUB 2100.

Converts X, galactic longitude or right ascension, and Y, galactic latitude or declination, into the corresponding P, the right ascension or galactic longitude, and Q, the declination or galactic latitude. All angles in *radians*.

Set SW(4) = +1 for equatorial to galactic conversion.

Set SW(4) = −1 for galactic to equatorial conversion.

Set FL(8) = 0 on first call. It is set to 1 by the routine and may be ignored thereafter.

Notes	Code
	2097 REM
	2098 REM Subroutine EQGAL
	2099 REM
2100 First call?	2100 IF FL(8)=1 THEN GOTO 2120
2105 Define galactic and other constants (radians)	2105 PI=3.1415926535 : TP=2.0*PI : GR=3.355395
	2110 GD=4.782202E-1 : AN=5.759587E-1
2115 We only need calculate these quantities once	2115 SL=SIN(GD) : CL=COS(GD) : FL(8)=1
2120 Calculate trigonometric values once	2120 CY=COS(Y) : SY=SIN(Y) : A=X-AN
2125 SW(4)=1 if equatorial to galactic conversion	2125 IF SW(4)=1 THEN A=X-GR
2130 Calculate more trigonometric values just once	2130 CA=COS(A) : SA=SIN(A) : B=SA
	2135 IF SW(4)=1 THEN B=CA
2140 FNS returns inverse sine	2140 S=(CY*CL*B)+(SY*SL) : Q=FNS(S)
	2145 IF SW(4)=-1 THEN GOTO 2155
2150 This line if equatorial to galactic . . .	2150 C=SY-(S*SL) : D=CY*SA*CL : A=AN : GOTO 2160
2155 . . . or this if galactic to equatorial	2155 C=CY*CA : D=(SY*CL)-(CY*SL*SA) : A=GR
2160 Make sure D is never zero	2160 IF ABS(D)<1E-20 THEN D=1E-20
	2165 P=ATN(C/D)+A
2170 Remove ambiguity of inverse tangent	2170 IF D<0 THEN P=P+PI
2175 Return P in range 0–2π radians	2175 IF P<0 THEN P=P+TP : GOTO 2175
	2180 IF P>TP THEN P=P-TP : GOTO 2180
2185 Return from EQGAL	2185 RETURN

```
1     REM
2     REM         Handling program HEQGAL
3     REM

5     DIM FL(20),SW(20)
```

10 Converts degrees to radians . . .
```
10    DEF FNM(W)=1.745329252E-2*W
```
15 . . . and radians to degrees
```
15    DEF FND(W)=5.729577951E1*W
```
20 FNS returns inverse sine
```
20    DEF FNS(W)=ATN(W/(SQR(1-W*W)+1E-20))

25    PRINT : PRINT : FL(8)=0
30    PRINT "Equatorial and galactic coords."
35    PRINT "------------------------------"
40    PRINT : PRINT
```

45 Which direction to convert?
50 Ask with YESNO
```
45    Q$="Equatorial to galactic (Y or N) .... "
50    PRINT : GOSUB 960
55    IF E=1 THEN GOTO 140
60    Q$="Galactic to equatorial (Y or N) .... "
65    GOSUB 960
```
70 Skip to the end if neither direction
```
70    IF E=0 THEN GOTO 205
```

75 Here for galactic to equatorial
80 Get the coordinates . . .
85 . . . and convert to radians with MINSEC then FNM
```
75    Q$="Galactic longitude (D,M,S) ........ "
80    PRINT : PRINT Q$; : INPUT XD,XM,XS
85    SW(1)=-1 : GOSUB 1000 : XA=FNM(X)
```
90 Now for the latitude . . .
```
90    Q$="Galactic latitude (D,M,S) ......... "
95    PRINT Q$; : INPUT XD,XM,XS
```
100 . . . in radians
```
100   SW(1)=-1 : GOSUB 1000 : Y=FNM(X) : X=XA
```
105 Set the switch, call EQGAL, and convert to degrees
```
105   SW(4)=-1 : GOSUB 2100 : P=FND(P) : Q=FND(Q)
```

110 Convert right scension to H,M,S form (MINSEC) . . .
```
110   X=P/15.0 : NC=9 : SW(1)=1 : GOSUB 1000
115   Q$="Right ascension (H,M,S) ........... "
```
120 . . . and display it neatly
125 Convert declination to D,M,S form . . .
```
120   PRINT Q$+"  "+MID$(OP$,4,12)
125   X=Q : GOSUB 1000
130   Q$="Declination (D,M,S) ............... "
```
135 . . . and display it too; skip to the end
```
135   PRINT Q$+OP$ : GOTO 205
```

140 Here for equatorial to galactic conversion
145 Get the coordinates . . .
150 . . . and convert to radians (MINSEC, FNM)
```
140   Q$="Right ascension (H,M,S) ........... "
145   PRINT : PRINT Q$; : INPUT XD,XM,XS
150   SW(1)=-1 : GOSUB 1000 : XA=FNM(X*15.0)
```
155 Now for the latitude . . .
```
155   Q$="Declination (D,M,S) ............... "
160   PRINT Q$; : INPUT XD,XM,XS
```
165 . . . also in radians
```
165   SW(1)=-1 : GOSUB 1000 : Y=FNM(X) : X=XA
```
170 Set the switch, call EQGAL, and convert to degrees
```
170   SW(4)=1 : GOSUB 2100 : P=FND(P) : Q=FND(Q)
```

175 Convert longitude to D,M,S form (MINSEC) . . .
```
175   X=P : NC=9 : SW(1)=1 : GOSUB 1000
180   Q$="Galactic longitude (D,M,S) ........ "
```
185 . . . and display it
190 Convert latitude to D,M,S form . . .
```
185   PRINT Q$+OP$
190   X=Q : GOSUB 1000
195   Q$="Galactic latitude (D,M,S) ......... "
```
200 . . . and display it
```
200   PRINT Q$+OP$
```

205 Another conversion?
210 Ask YESNO for the answer
```
205   Q$="Again (Y or N) .................... "
210   PRINT : GOSUB 960
215   IF E=1 THEN GOTO 45
220   STOP

INCLUDE YESNO, MINSEC, EQGAL
```

2100 EQGCL

Example

```
Equatorial and galactic coords.
-------------------------------


Equatorial to galactic (Y or N) .... ? Y

Right ascension (H,M,S) ............ ? 19,19,00
Declination (D,M,S) ............... ? 66,30,00
Galactic longitude (D,M,S) ........   + 97 44 50.17
Galactic latitude (D,M,S) .........   + 22  4 30.54

Again (Y or N) .................... ? Y

Equatorial to galactic (Y or N) .... ? N
Galactic to equatorial (Y or N) .... ? Y

Galactic longitude (D,M,S) ........ ? 97,44,50.17
Galactic latitude (D,M,S) ......... ? 22,4,30.54
Right ascension (H,M,S) ...........    19 19  0.00
Declination (D,M,S) ...............   + 66 30  0.00

Again (Y or N) .................... ? Y

Equatorial to galactic (Y or N) .... ? Y

Right ascension (H,M,S) ........... ? 12,0,0
Declination (D,M,S) ............... ? 88,33,0
Galactic longitude (D,M,S) ........   +123 21  3.99
Galactic latitude (D,M,S) .........   + 28 48 59.56

Again (Y or N) .................... ? Y

Equatorial to galactic (Y or N) .... ? N
Galactic to equatorial (Y or N) .... ? Y

Galactic longitude (D,M,S) ........ ? 123,21,3.99
Galactic latitude (D,M,S) ......... ? 28,48,59.56
Right ascension (H,M,S) ...........    11 59 59.99
Declination (D,M,S) ...............   + 88 33  0.00

Again (Y or N) .................... ? N
```

2200 GENCON

This routine converts any one of the coordinate pairs azimuth and altitude, hour angle and declination, right ascension and declination, ecliptic longitude and ecliptic latitude, galactic longitude and galactic latitude, into any other pair.

The subroutines described earlier in this book for converting between the various coordinate systems are quite satisfactory for normal use where you have a requirement for one particular conversion. However, it is sometimes convenient to have a more general routine to hand which can be used to convert from one system to any other system, at the flick of a (software) switch. We can use matrices for this purpose, ordered sets of numbers set out in rows and columns. The manipulation of the matrices is always the same; to change from one system to another you merely have to change the numbers in the matrices.

Subroutine GENCON is able to produce and manipulate just four different matrices, appropriate for conversions between azimuth/altitude and hour angle/declination, hour angle/declination and right ascension/declination, right ascension/declination and ecliptic longitude/latitude, and right ascension/ declination and galactic longitude/latitude. However, you can use it to convert coordinates from any of the systems to any other system by serial conversion. For example, suppose that you wish to know the azimuth and altitude of a pulsar whose galactic coordinates you have to hand. You would first convert to right ascension/declination, then to hour angle/declination, and finally to azimuth/altitude. GENCON will do this triple conversion automatically.

The coordinates to be converted are input to GENCON as usual via the variables X and Y (radians). These might represent the galactic longitude and latitude, for example. The converted coordinate pair (e.g. the azimuth and altitude) are returned in the variables P and Q (also in radians). You must specify the conversions to be carried out with the three-element control array CC. Each element must be set to an integer in the range 0–8 inclusive. The array is read in the order CC(1), CC(2), CC(3) by the routine, and the integer obtained used to direct the subsequent action. When GENCON encounters the value 0 is returns control to the calling program with the results in P and Q. Other values produce conversions as laid out in the list below. Thus, to convert from galactic coordinates to horizon coordinates, you would set X to the galactic longitude, Y to the galactic latitude, CC(1) to 8, CC(2) to 4 and CC(3) to 2 before calling the routine.

Other values that need to be specified before calling GENCON depend on the conversions to be carried out. Where azimuths and altitudes are involved you need the geographical latitude, GP, in radians. Hour angle/declination to right ascension/declination needs the local sidereal time, ST, in radians. Finally, converting from equatorial to ecliptic coordinates depends on the obliquity of the ecliptic. GENCON calls OBLIQ to find this, and therefore needs the date set in DY, MN, YR.

Flags used by GENCON include FL(4) to indicate that the geographical latitude has not changed since the last call, FL(7) to indicate no change of date, and FL(9) to indicate no change of sidereal time. Set these to 0 for the first call, or whenever the corresponding parameter changes between calls. Otherwise, you can leave them alone.

The handling program, HGENCON, illustrates the use of the subroutine, but at the same time provides a general purpose conversion program. It incorporates various interlocks to guard against operator error, but check its operation thoroughly before relying on its results to ensure that there are no copying errors. Please let me know of any bugs you find in the logic!

Details of GENCON

Called by GOSUB 2200.

Converts coordinates X and Y specified in one system to coordinates P and Q specified in another system according to the setting of the three-element control array CC. All angles in *radians*.

Set

CC(J) = 0 to end the conversion

CC(J) = 1 for azimuth/altitude to hour angle/declination

CC(J) = 2 for hour angle/declination to azimuth/altitude

CC(J) = 3 for hour angle/declination to right ascension/declination

CC(J) = 4 for right ascension/declination to hour angle/declination

CC(J) = 5 for right ascension/declination to ecliptic longitude/latitude

CC(J) = 6 for ecliptic longitude/latitude to right ascension/declination

CC(J) = 7 for right ascension/declination to galactic longitude/latitude

CC(J) = 8 for galactic longitude/latitude to right ascension/declination

where J = 1, 2, or 3. If CC(1) = 0, control is returned immediately to the calling program without alteration to P and Q.

If any of CC(J) = 1 or 2: set GP equal to the geographical latitude in radians and FL(4) to 0 before calling the routine for the first time with a new value of GP.

If any of CC(J) = 3 or 4: set ST equal to the local sidereal time in radians and FL(9) to 0 before calling the routine for the first time with a new value of ST. If any of CC(J) = 5 or 6: enter the date as usual via the variables DY (day), MN (month), YR (year) and set FL(7) to 0 before calling the routine for the first time for a new date.

Arrays CC(3), CV(3), HL(3), and MT(3,3) must be declared before calling GENCON.

Other routine called: OBLIQ (1700).

Notes	Code
	2197 REM
	2198 REM Subroutine GENCON
	2199 REM
2200 Any conversions at all?	2200 IF CC(1)=0 THEN RETURN
	2214 REM Convert input into column vector
2215 Set up the three-element column vector . . .	2215 CN=COS(Y) : CV(1)=COS(X)*CN
2220 . . . corresponding to the input coordinates X,Y	2220 CV(2)=SIN(X)*CN : CV(3)=SIN(Y)
2224 *** the control loop ***	2224 REM Construct the matrices
2225 Control loop, II; do whatever CC(II) commands . . .	2225 PI=3.1415926536 : TP=2.0*PI : FOR II=1 TO 3
2230 . . . no more conversions	2230 IF CC(II)=0 THEN GOTO 2295
2235 . . . horizon to hour angle/declination	2235 IF CC(II)=1 THEN GOSUB 2325 : GOTO 2275
2240 . . . hour angle/declination to horizon	2240 IF CC(II)=2 THEN GOSUB 2325 : GOTO 2275
2245 . . . hour angle/declination to right ascension/declination	2245 IF CC(II)=3 THEN GOSUB 2360 : GOTO 2275
2250 . . . right ascension/declination to hour angle/declination	2250 IF CC(II)=4 THEN GOSUB 2360 : GOTO 2275
2255 . . . right ascension/declination to ecliptic	2255 IF CC(II)=5 THEN GOSUB 2395 : GOTO 2275
2260 . . . ecliptic to right ascension/declination	2260 IF CC(II)=6 THEN GOSUB 2430 : GOTO 2275
2265 . . . right ascension/declination to galactic	2265 IF CC(II)=7 THEN GOSUB 2445 : GOTO 2275
2270 . . . galactic to right ascension/ declination	2270 IF CC(II)=8 THEN GOSUB 2470
	2274 REM Multiply each matrix by the column vector
2275 Multiply the column vector CV by the matrix MT . . .	2275 FOR J=1 TO 3 : SM=0.0
	2280 FOR I=1 TO 3 : SM=SM+MT(I,J)*CV(I) : NEXT I
	2285 HL(J)=SM : NEXT J : FOR I=1 TO 3 : CV(I)=HL(I)
2290 . . . putting the result back into CV	2290 NEXT I : NEXT II : II=4
2294 *** end of control loop ***	2294 REM Convert column vector into output
2295 Calculate P and Q from the elements of CV	2295 IF ABS(CV(1))<1E-20 THEN CV(1)=1E-20
	2300 P=ATN(CV(2)/CV(1)) : Q=FNS(CV(3))
2305 Allow for ambiguity of inverse tangent	2305 IF CV(1)<0 THEN P=P+PI
2310 Make sure P is in range 0–2π	2310 IF P<0 THEN P=P+TP : GOTO 2310
	2315 IF P>TP THEN P=P-TP : GOTO 2315
2320 Normal return from GENCON	2320 RETURN
2324 Set up MT(3,3) for azimuth, altitude and hour angle/declination	2324 REM Az/Alt and Ha/Dec
2325 New geographical latitude?	2325 IF FL(4)=1 THEN GOTO 2335
2330 Calculate these quantities only once	2330 SF=SIN(GP) : CF=COS(GP) : FL(4)=1
2335 Some elements are 0 . . .	2335 FOR J=1 TO 3 : FOR I=1 TO 3
	2340 MT(I,J)=0.0 : NEXT I : NEXT J
2345 . . . others involve the geographical latitude	2345 MT(1,1)=-SF : MT(3,1)=CF
	2350 MT(2,2)=-1.0 : MT(1,3)=CF
2355 MT is the same for conversion in either direction	2355 MT(3,3)=SF : RETURN
2359 Set up MT(3,3) for right ascension, declination and hour angle, declination	2359 REM Ra/Dec and Ha/Dec

2360	New local sidereal time?	2360	`IF FL(9)=1 THEN GOTO 2370`
2365	Calculate these quantities once only	2365	`SS=SIN(ST) : CS=COS(ST) : FL(9)=1`
2370	Some elements are 0 . . .	2370	`FOR J=1 TO 3 : FOR I=1 TO 3`
		2375	`MT(I,J)=0.0 : NEXT I : NEXT J`
2380	. . . others involve the local sidereal time	2380	`MT(1,1)=CS : MT(2,1)=SS`
		2385	`MT(1,2)=SS : MT(2,2)=-CS`
2390	MT is the same for conversion in either direction	2390	`MT(3,3)=1.0 : RETURN`
2394	Set up MT(3,3) for equatorial and ecliptic coordinates	2394	`REM Ra/Dec and Lam/Bet`
2395	New date?	2395	`IF FL(7)=1 THEN GOTO 2410`
2400	Call OBLIQ to find obliquity of the ecliptic	2400	`FL(5)=0 : GOSUB 1700 : FL(7)=1`
2405	Calculate these values once only	2405	`E=FNM(OB) : SE=SIN(E) : CE=COS(E)`
2410	Some elements of MT are 0 . . .	2410	`FOR J=1 TO 3 : FOR I=1 TO 3`
		2415	`MT(I,J)=0.0 : NEXT I : NEXT J`
2420	. . . others involve the obliquity.	2420	`MT(1,1)=1.0 : MT(2,2)=CE : MT(3,2)=SE`
2425	This for right ascension/declination to ecliptic conversion	2425	`MT(2,3)=-SE : MT(3,3)=CE : RETURN`
2430	This for ecliptic to right ascension/declination; set MT as above . . .	2430	`GOSUB 2395 : A=MT(3,2)`
2435	. . . and then just transpose two elements	2435	`MT(3,2)=MT(2,3) : MT(2,3)=A`
		2440	`RETURN`
2444	Set up MT(3,3) for equatorial and galactic coordinates	2444	`REM Galactic coordinates and Ra/Dec`
2445	Specify every element as a number (they never change)	2445	`MT(1,1)=-0.0669887 : MT(2,1)=-0.8727558`
2450	. . . this for right ascension/declination to galactic	2450	`MT(3,1)=-0.4835389 : MT(1,2)=0.4927285`
		2455	`MT(2,2)=-0.450347 : MT(3,2)=0.7445846`
		2460	`MT(1,3)=-0.8676008 : MT(2,3)=-0.1883746`
		2465	`MT(3,3)=0.4601998 : RETURN`
2470	. . . this for galactic to right ascension/declination	2470	`MT(1,1)=-0.0669887 : MT(1,2)=-0.8727558`
		2475	`MT(1,3)=-0.4835389 : MT(2,1)=0.4927285`
		2480	`MT(2,2)=-0.450347 : MT(2,3)=0.7445846`
		2485	`MT(3,1)=-0.8676008 : MT(3,2)=-0.1883746`
		2490	`MT(3,3)=0.4601998 : RETURN`

```
1       REM
2       REM          Handling program HGENCON
3       REM
```

5 Declare these arrays for the sub-routines . . .

```
5       DIM FL(20),ER(20),SW(20),MT(3,3),CV(3),HL(3),CC(3)
```

10 . . . and these for the handling program
15 Needed by JULDAY
20 Degrees to radians . . .
25 . . . and radians to degrees conversion
30 FNS returns the inverse sine

```
10      DIM CD$(8),CE$(8),CK(8,3),HS(4)
15      DEF FNI(W)=SGN(W)*INT(ABS(W))
20      DEF FNM(W)=1.745329252E-2*W
25      DEF FND(W)=5.729577951E1*W
30      DEF FNS(W)=ATN(W/(SQR(1-W*W)+1E-20))
```

35 CK is the array which checks to see that the . . .
40 . . . combinations of settings of the control array . . .
45 . . . CC are allowed (e.g. 1,6,0 is not allowed)

```
35      CK(1,1)=2 : CK(1,2)=3 : CK(2,1)=1 : CK(3,1)=4
40      CK(3,2)=5 : CK(3,3)=7 : CK(4,1)=2 : CK(4,2)=3
45      CK(5,1)=6 : CK(6,1)=7 : CK(6,2)=5 : CK(6,3)=4
50      CK(7,1)=8 : CK(8,1)=5 : CK(8,2)=4 : CK(8,3)=7
```

```
55      PRINT : PRINT : SP=0
60      PRINT "General coordinate conversions"
65      PRINT "-----------------------------"
70      PRINT : PRINT
```

75 What is your desire? . . .
80 . . . horizon to hour angle/declination?
85 . . . or the other way round?
90 . . . hour angle/declination to right ascension/declination?
95 . . . or the other way round?
100 . . . equatorial to ecliptic longitude and latitude?
105 . . . or the other way round?
110 . . . equatorial to galactic longitude and latitude?
115 . . . or the other way round?

```
75      PRINT "Conversion codes:" : PRINT
80      CD$(1)="azi,alt  >  ha ,dec" : CE$(1)="   ...code 1"
85      CD$(2)="azi,alt  <  ha ,dec" : CE$(2)="   ...code 2"
90      CD$(3)="ha ,dec  >  ra ,dec" : CE$(3)="   ...code 3"
95      CD$(4)="ha ,dec  <  ra, dec" : CE$(4)="   ...code 4"
100     CD$(5)="ra ,dec  >  lam,bet" : CE$(5)="   ...code 5"
105     CD$(6)="ra ,dec  <  lam,bet" : CE$(6)="   ...code 6"
110     CD$(7)="ra ,dec  >   l,b  "  : CE$(7)="   ...code 7"
115     CD$(8)="ra ,dec  <   l,b  "  : CE$(8)="   ...code 8"
```

120 Display the options . . .
125 . . . plus this one of course . . .
130 . . . and get the operator's choices
135 The first choice . . .

```
120     FOR I=1 TO 8 : PRINT CD$(I)+CE$(I) : NEXT I
125     PRINT "No more conversions ...code 0"
130     PRINT : PRINT "Please input up to 3 codes"
135     Q$="First conversion code (0-8) ...... "
140     PRINT Q$; : INPUT I : CC(1)=I
```

145 . . . and make sure that it is valid
150 If 0 we don't wish to convert anything
155 Get the second choice . . .

```
145     IF I>8 OR I<0 THEN GOTO 140
150     IF I=0 THEN GOTO 395
155     Q$="Second conversion code (0-8) ..... "
160     PRINT Q$; : INPUT I : CC(2)=I
```

165 . . . and make sure that it is valid . . .
170 If 0, one conversion is sufficient
175 . . . and then check to see that it can follow choice 1

```
165     IF I>8 OR I<0 THEN GOTO 160
170     IF I=0 THEN GOTO 225
175     J=CC(1) : FOR K=1 TO 3
```

180 Jump if choice is allowed . . .
185 . . . or deliver a raspberry if it isn't!
190 Get the third choice . . .

```
180     IF I=CK(J,K) THEN GOTO 190
185     NEXT K : PRINT "** impossible" : GOTO 160
190     Q$="Final conversion code (0-8) ...... "
```

200 . . . and make sure that it is valid . . .
205 If 0, two conversions are sufficient
210 . . . and then check to see that it can follow choice 2
215 Jump if allowed . . .
220 . . . or complain if it isn't

```
195     PRINT Q$; : INPUT I : CC(3)=I
200     IF I>8 OR I<0 THEN GOTO 195
205     IF I=0 THEN GOTO 225
210     J=CC(2) : FOR K=1 TO 3
215     IF I=CK(J,K) THEN GOTO 225
220     NEXT K : PRINT "** impossible" : GOTO 195
```

225 Display all three selections . . .

```
225     PRINT : PRINT "Your selection is:"
230     FOR I=1 TO 3
```

```
                                    235   IF CC(I)=0 THEN GOTO 245
                                    240   PRINT CD$(CC(I)) : NEXT I
245  ...and ask if that is really what you mean   245   Q$="Is this OK (Y or N) ............. "
     to do
250  Get the answer with YESNO        250   PRINT : GOSUB 960 : PRINT
255  Back to the drawing-board if it isn't   255   IF E=0 THEN GOTO 70

260  Get the input coordinates        260   IF CC(1)=1 THEN GOSUB 415 : GOTO 300
265  The type depends on the first choice...   265   IF CC(1)=2 THEN GOSUB 430 : MU=MU*15.0 : GOTO 300
270  ...so call the appropriate input  270   IF CC(1)=3 THEN GOSUB 430 : MU=MU*15.0 : GOTO 300
     routine...
275  ...remembering to convert hours to   275   IF CC(1)=4 THEN GOSUB 445 : MU=MU*15.0 : GOTO 300
     degrees if needed                280   IF CC(1)=5 THEN GOSUB 445 : MU=MU*15.0 : GOTO 300
                                    285   IF CC(1)=6 THEN GOSUB 460 : GOTO 300
                                    290   IF CC(1)=7 THEN GOSUB 445 : MU=MU*15.0 : GOTO 300
                                    295   IF CC(1)=8 THEN GOSUB 475

300  Now we have to get the other bits and   300   PRINT : FOR I=1 TO 3
     pieces...
305  ...such as geographical latitude...   305   IF CC(I)=0 THEN GOTO 345
310  ...local sidereal time etc. What we need   310   IF CC(I)=1 THEN GOSUB 505 : GOTO 340
     depends...
315  ...on the selection of conversion types   315   IF CC(I)=2 THEN GOSUB 505 : GOTO 340
                                    320   IF CC(I)=3 THEN GOSUB 545 : GOTO 340
                                    325   IF CC(I)=4 THEN GOSUB 545 : GOTO 340
                                    330   IF CC(I)=5 THEN GOSUB 750 : GOTO 340
                                    335   IF CC(I)=6 THEN GOSUB 750
                                    340   NEXT I

345  Call GENCON to do the business   345   X=MU : Y=NU : GOSUB 2200 : PRINT : I=II-1
350  This should never occur          350   IF I=0 THEN GOTO 395

355  I now points to the last conversion. Use   355   IF CC(I)=1 THEN GOSUB 790 : GOTO 395
     CC(I)...
360  ...to determine what sort of coordinates   360   IF CC(I)=2 THEN GOSUB 775 : GOTO 395
     P,Q are...
365  ...and call the appropriate display   365   IF CC(I)=3 THEN GOSUB 805 : GOTO 395
     routine                          370   IF CC(I)=4 THEN GOSUB 790 : GOTO 395
                                    375   IF CC(I)=5 THEN GOSUB 820 : GOTO 395
                                    380   IF CC(I)=6 THEN GOSUB 805 : GOTO 395
                                    385   IF CC(I)=7 THEN GOSUB 835 : GOTO 395
                                    390   IF CC(I)=8 THEN GOSUB 805

395  Another conversion?              395   Q$="Again (Y or N) .................. "
400  Use YESNO to see                 400   PRINT : GOSUB 960
                                    405   IF E=0 THEN STOP
410  Reset the local switches (array HS) and   410   HS(1)=0 : HS(2)=0 : HS(3)=0 : HS(4)=0 : GOTO 70
     jump to start

415  Messages for input...            415   Q1$="Azimuth (D,M,S) ................ "
                                    420   Q2$="Altitude (D,M,S) ................ "
                                    425   GOTO 485
                                    430   Q1$="Hour angle (H,M,S) .............. "
                                    435   Q2$="Declination (D,M,S) ............. "
                                    440   GOTO 485
                                    445   Q1$="Right ascension (H,M,S) ......... "
                                    450   Q2$="Declination (D,M,S) ............. "
                                    455   GOTO 485
                                    460   Q1$="Ecliptic longitude (D,M,S) ...... "
                                    465   Q2$="Ecliptic latitude  (D,M,S) ...... "
                                    470   GOTO 485
                                    475   Q1$="Galactic longitude (D,M,S) ...... "
                                    480   Q2$="Galactic latitude  (D,M,S) ...... "
485  Display the message, and get the   485   PRINT Q1$; : INPUT XD,XM,XS
     parameters
```

490	Convert to decimal form with MINSEC and to radians	490	`SW(1)=-1 : GOSUB 1000 : MU=FNM(X)`
		495	`PRINT Q2$; : INPUT XD,XM,XS`
		500	`GOSUB 1000 : NU=FNM(X) : RETURN`
505	Geographical latitude input; do we need a new one?	505	`IF HS(2)=1 THEN RETURN`
510	Have we already got an old one?	510	`IF FL(4)=0 THEN GOTO 530`
		515	`Q$="New latitude (Y or N) .......... "`
520	Do we want a new one?	520	`GOSUB 960 : HS(2)=1`
525	Jump if no . . .	525	`IF E=0 THEN RETURN`
530	. . . otherwise get a new latitude . . .	530	`Q$="Geographical latitude (D,M,S) .... "`
		535	`PRINT Q$; : INPUT XD,XM,XS : SW(1)=-1 : FL(4)=0`
540	. . . in radians (MINSEC and FNM)	540	`GOSUB 1000 : GP=FNM(X) : HS(2)=1 : RETURN`
545	Time input; do we need a new local sidereal time?	545	`IF HS(1)=1 THEN RETURN`
550	Have we an old one already?	550	`IF FL(9)=0 THEN GOTO 570`
555	Do we want a new one anyway?	555	`Q$="New local sidereal time (Y or N)   "`
560	Call YESNO to see	560	`GOSUB 960 : HS(1)=1`
565	Jump if the answer is no	565	`IF E=0 THEN RETURN`
570	Perhaps we know the local sidereal time?	570	`Q$="Do you know the LST (Y or N) ..... "`
575	Call YESNO to find out	575	`GOSUB 960 : HS(1)=1 : FL(9)=0`
		580	`IF E=0 THEN GOTO 600`
585	Ask for it if the answer is yes . . .	585	`Q$="Local sidereal time (H,M,S) ...... "`
		590	`PRINT Q$; : INPUT XD,XM,XS : SW(1)=-1`
595	. . . and convert it to radians	595	`GOSUB 1000 : ST=FNM(X*15.0) : RETURN`
600	SP=1 signifies that the location has not changed	600	`IF SP=0 THEN GOTO 620`
605	Do we wish to change the location?	605	`Q$="New location (Y or N) ........... "`
610	Ask with YESNO	610	`GOSUB 960`
615	Jump if the answer is no	615	`IF E=0 THEN GOTO 655`
620	Now we have to get everything we need to know . . .	620	`Q$="Daylight saving (H ahead of zone)  "`
625	. . . so that we can use TIME to calculate the . . .	625	`PRINT Q$; : INPUT DS : SP=1`
630	. . . local sidereal time from the local civil time	630	`Q$="Time zone (hours; West negative)   "`
		635	`PRINT Q$; : INPUT TZ`
		640	`Q$="Geog. longitude (D,M,S; W neg.) .. "`
		645	`PRINT Q$; : INPUT XD,XM,XS`
		650	`SW(1)=-1 : GOSUB 1000 : GL=X`
655	We need the date (but it might already be known)	655	`GOSUB 710`
		660	`Q$="Local civil time (H,M,S) ......... "`
		665	`PRINT Q$; : INPUT XD,XM,XS : SW(1)=-1`
670	Call TIME to find the local sidereal time	670	`GOSUB 1000 : TM=X : SW(2)=1 : GOSUB 1300`
675	Was the date impossible?	675	`IF ER(1)=1 THEN GOTO 395`
680	Display all we know about the times	680	`Q$="Universal time (H,M,S) ........... "`
		685	`PRINT Q$+UT$`
		690	`Q$="Greenwich sidereal time (H,M,S) .. "`
		695	`PRINT Q$+SG$`
		700	`Q$="Local sidereal time (H,M,S) ...... "`
705	Converts the local sidereal time to radians	705	`PRINT Q$+TL$ : ST=FNM(TL*15.0) : RETURN`
710	HS(3)=1 if the date is already known	710	`IF HS(3)=1 THEN RETURN`
715	Do we know an old date?	715	`IF FL(1)=0 THEN PRINT : GOTO 735`
720	Do we want a new date?	720	`Q$="New calendar date (Y or N) ....... "`
725	YESNO returns with the answer	725	`PRINT : GOSUB 960 : HS(3)=1`
730	Jump if no	730	`IF E=0 THEN RETURN`
735	Get a new date . . .	735	`Q$="Calendar date (D,M,Y) ........... "`
		740	`PRINT Q$; : INPUT DY,MN,YR`
745	. . . and reset the flags to let the sub-routines know	745	`FL(3)=0 : FL(7)=0 : RETURN`

750	HS(4)=1 signifies that we already know the obliquity	750	`IF HS(4)=1 THEN RETURN`
755	If we don't, must we add in the nutation?	755	`Q$="Include nutation (Y or N) ........ "`
760	Call YESNO to see	760	`GOSUB 960 : FL(6)=1 : HS(4)=1`
765	FL(6)=0 to include nutation	765	`IF E=1 THEN FL(6)=0 : FL(7)=0`
770	Get a date, and call OBLIQ for a new obliquity	770	`GOSUB 710 : GOSUB 1700 : RETURN`
775	Output messages . . .	775	`Q1$="Azimuth (D M S) .................. "`
		780	`Q2$="Altitude (D M S) ................ "`
785	I signifies whether the output is HMS or DMS	785	`I=0 : GOTO 850`
		790	`Q1$="Hour angle (H M S) .............. "`
		795	`Q2$="Declination (D M S) ............. "`
		800	`I=1 : P=P/15.0 : GOTO 850`
		805	`Q1$="Right ascension (H M S) ......... "`
		810	`Q2$="Declination (D M S) ............. "`
		815	`I=1 : P=P/15.0 : GOTO 850`
		820	`Q1$="Ecliptic longitude (D M S) ....... "`
		825	`Q2$="Ecliptic latitude  (D M S) ....... "`
		830	`I=0 : GOTO 850`
		835	`Q1$="Galactic longitude (D M S) ....... "`
		840	`Q2$="Galactic latitude  (D M S) ....... "`
		845	`I=0`
850	Convert to degrees and then to minutes and seconds form	850	`X=FND(P) : SW(1)=1 : GOSUB 1000`
855	Truncate the sign and line up the columns if HMS	855	`IF I=1 THEN OP$="    "+MID$(OP$,4,12)`
		860	`PRINT Q1$+OP$`
		865	`X=FND(Q) : SW(1)=1 : GOSUB 1000`
		870	`PRINT Q2$+OP$ : RETURN`

```
INCLUDE YESNO, MINSEC, JULDAY, TIME, OBLIQ, NUTAT, GENCON
```

2200 GENCON

Example

```
General coordinate conversions
------------------------------


Conversion codes:

azi,alt  >  ha ,dec   ...code 1
azi,alt  <  ha ,dec   ...code 2
ha ,dec  >  ra ,dec   ...code 3
ha ,dec  <  ra, dec   ...code 4
ra ,dec  >  lam,bet   ...code 5
ra ,dec  <  lam,bet   ...code 6
ra ,dec  >   l,b      ...code 7
ra ,dec  <   l,b      ...code 8
No more conversions ...code 0


Please input up to 3 codes
First conversion code (0-8) ...... ? 4
Second conversion code (0-8) ..... ? 2
Final conversion code (0-8) ...... ? 0


Your selection is:
ha ,dec  <  ra, dec
azi,alt  <  ha ,dec

Is this OK (Y or N) .............. ? Y

Right ascension (H,M,S) .......... ? 20,40,5.2
Declination (D,M,S) .............. ? -22,12,0

Do you know the LST (Y or N) ..... ? N
Daylight saving (H ahead of zone)  ? 0
Time zone (hours; West negative)   ? -5
Geog. longitude (D,M,S; W neg.) .. ? -77,0,0

Calendar date (D,M,Y) ............ ? 1,2,1984
Local civil time (H,M,S) ......... ? 7,23,0
Universal time (H,M,S) ...........   12 23  0.00
Greenwich sidereal time (H,M,S) ..   21  6 37.98
Local sidereal time (H,M,S) ......   15 58 37.98
Geographical latitude (D,M,S) .... ? 38,55,0

Azimuth (D M S) ..................   +119 18 14.59
Altitude (D M S) .................   +  0 16 15.88

Again (Y or N) ................... ? Y


Conversion codes:

azi,alt  >  ha ,dec   ...code 1
azi,alt  <  ha ,dec   ...code 2
ha ,dec  >  ra ,dec   ...code 3
ha ,dec  <  ra, dec   ...code 4
ra ,dec  >  lam,bet   ...code 5
ra ,dec  <  lam,bet   ...code 6
ra ,dec  >   l,b      ...code 7
ra ,dec  <   l,b      ...code 8
No more conversions ...code 0

Please input up to 3 codes
First conversion code (0-8) ...... ? 1
Second conversion code (0-8) ..... ? 3
Final conversion code (0-8) ...... ? 0
```

```
Your selection is:
azi,alt  > ha ,dec
ha ,dec  > ra ,dec

Is this OK (Y or N) .............. ? Y

Azimuth (D,M,S) .................. ? 119,18,14.59
Altitude (D,M,S) ................. ? 0,16,15.88

New latitude (Y or N) ........... ? N
New local sidereal time (Y or N)   ? N

Right ascension (H M S) ..........    20 40  5.20
Declination (D M S) ..............   - 22 12  0.00

Again (Y or N) ................... ? Y


Conversion codes:

azi,alt  > ha ,dec   ...code 1
azi,alt  < ha ,dec   ...code 2
ha ,dec  > ra ,dec   ...code 3
ha ,dec  < ra, dec   ...code 4
ra ,dec  > lam,bet   ...code 5
ra ,dec  < lam,bet   ...code 6
ra ,dec  >   l,b     ...code 7
ra ,dec  <   l,b     ...code 8
No more conversions ...code 0

Please input up to 3 codes
First conversion code (0-8) ...... ? 8
Second conversion code (0-8) ..... ? 3
** impossible
Second conversion code (0-8) ..... ? 4
Final conversion code (0-8) ...... ? 1
** impossible
Final conversion code (0-8) ...... ? 2

Your selection is:
ra ,dec  <   l,b
ha ,dec  <  ra, dec
azi,alt  < ha ,dec

Is this OK (Y or N) .............. ? Y

Galactic longitude (D,M,S) ....... ? 97,44,50.17
Galactic latitude  (D,M,S) ....... ? 22,4,30.54

New local sidereal time (Y or N)   ? Y
Do you know the LST (Y or N) ..... ? N
New location (Y or N) ........... ? Y
Daylight saving (H ahead of zone)  ? 0
Time zone (hours; West negative)   ? 0
Geog. longitude (D,M,S; W neg.) .. ? 0,2,5.67

New calendar date (Y or N) ....... ? Y
Calendar date (D,M,Y) ............ ? 23,11,1991
Local civil time (H,M,S) ......... ? 14,29,20
Universal time (H,M,S) ...........   14 29 20.00
Greenwich sidereal time (H,M,S) ..   18 37 34.64
Local sidereal time (H,M,S) ......   18 37 43.01
New latitude (Y or N) ............ ? Y
Geographical latitude (D,M,S) .... ? 52,10,12
```

2200 GENCON

```
Azimuth (D M S) ...................    + 15 47 30.11
Altitude (D M S) ................       + 74 46 51.43

Again (Y or N) ................... ? Y


Conversion codes:

azi,alt  >  ha ,dec   ...code 1
azi,alt  <  ha ,dec   ...code 2
ha ,dec  >  ra ,dec   ...code 3
ha ,dec  <  ra, dec   ...code 4
ra ,dec  >  lam,bet   ...code 5
ra ,dec  <  lam,bet   ...code 6
ra ,dec  >    l,b     ...code 7
ra ,dec  <    l,b     ...code 8
No more conversions ...code 0


Please input up to 3 codes
First conversion code (0-8) ...... ? 1
Second conversion code (0-8) ..... ? 3
Final conversion code (0-8) ...... ? 7


Your selection is:
azi,alt  >  ha ,dec
ha ,dec  >  ra ,dec
ra ,dec  >    l,b

Is this OK (Y or N) .............. ? Y

Azimuth (D,M,S) .................. ? 15,47,30.11
Altitude (D,M,S) ................. ? 74,46,51.43

New latitude (Y or N) ............ ? N
New local sidereal time (Y or N)   ? N

Galactic longitude (D M S) .......    + 97 44 50.16
Galactic latitude  (D M S) .......    + 22  4 30.54

Again (Y or N) ................... ? Y


Conversion codes:

azi,alt  >  ha ,dec   ...code 1
azi,alt  <  ha ,dec   ...code 2
ha ,dec  >  ra ,dec   ...code 3
ha ,dec  <  ra, dec   ...code 4
ra ,dec  >  lam,bet   ...code 5
ra ,dec  <  lam,bet   ...code 6
ra ,dec  >    l,b     ...code 7
ra ,dec  <    l,b     ...code 8
No more conversions ...code 0


Please input up to 3 codes
First conversion code (0-8) ...... ? 8
Second conversion code (0-8) ..... ? 0

Your selection is:
ra ,dec  <    l,b

Is this OK (Y or N) .............. ? Y

Galactic longitude (D,M,S) ....... ? 97,44,50.16
Galactic latitude  (D,M,S) ....... ? 22,4,30.54
```

```
Right ascension (H M S) ..........      19 19  0.01
Declination (D M S) ..............    + 66 29 59.98

Again (Y or N) .................. ? N
```

2500 PRCESS1
2600 PRCESS2

These routines transform equatorial coordinates from one epoch to another, making allowance for precession. PRCESS1 uses an approximate method. PRCESS2 performs the transformation rigorously.

The right ascension and declination of a celestial body change slowly with time because of *luni–solar precession*, a gyrating motion of the Earth's axis caused by the gravitational effects of the Sun and the Moon on the equatorial bulge of the Earth. The Earth's axis describes a cone about a line perpendicular to the ecliptic through the centre of the Earth with half-angle approximately 23.5 degrees, completing one circuit in about 25 800 years. The change is thus about 50 arcseconds per year and this must be allowed for when comparing equatorial coordinates at two different epochs.

PRCESS1

Two subroutines are given here which will correct equatorial coordinates for the effects of precession between one epoch and another. The first of these, PRCESS1 (2500), uses an approximate method which is simple and compact, but which nevertheless gives good results. You should find it sufficient for most purposes. The corrections, X1 and Y1, which have to be made to the right ascension, X, and declination, Y, measured at epoch A, to find the corresponding values at epoch B, are well approximated by the expressions:

$$X1 = (MP + (NP \times SIN(X) \times TAN(Y))/15) \times NY$$
$$Y1 = NP \times COS(X) \times NY,$$

where NY is the number of years between the two epochs, and MP and NP are 'constants' which change slowly with time. The routine given here makes use of these expressions, calculating MP and NP at each of the two epochs and taking the average. It will not work for regions which are very close to the north and south celestial poles where the value of TAN(Y) approaches infinity and rapidly exceeds the maximum number which your machine can handle. The dates of the two epochs are input as DA, MA, YA for epoch 1 and DB, MB, YB for epoch 2. PRCESS1 makes calls to JULDAY (1100) to calculate the number of days between them.

Execution of PRCESS1 is controlled by the flag FL(10). It is obviously only necessary to calculate the constants once for any pair of epochs and this is done in lines 2505–2560 when FL(10) = 0. The flag is set to 1 in line 2510, and subsequent calls with FL(10) = 1 jump straight to line 2565. Therefore, set this flag to 0 when calling the routine for the first time or whenever either or both of the epochs is changed. Otherwise, leave the flag set at 1 so that execution time is reduced. The routine itself takes care of FL(1) when calling JULDAY.

Formulae for PRCESS1

$X = X + X1$

$Y = Y + Y1$

$X1 = (MP + (NP \times SIN(X) \times TAN(Y)/15)) \times 7.272205E{-}5 \times NY$

$Y1 = NP \times COS(X) \times 4.848137E{-}6 \times NY$

$MP = (M1 + M2)/2$

$NP = (N1 + N2)/2$

$M1 = 3.07234 + (1.86E{-}3 \times T1)$

$M2 = 3.07234 + (1.86E{-}3 \times T2)$

$N1 = 20.0468 - (8.5E{-}3 \times T1)$

$N2 = 20.0468 - (8.5E{-}3 \times T2)$

$T1 = DK/36525$

$T2 = DJ/36525$

$NY = (DJ - DK)/365.2425$

DJ = Julian days since 1900 January 0.5 to epoch 2

DK = Julian days since 1900 January 0.5 to epoch 1

Details of PRCESS1

Called by GOSUB 2500.

Corrects the right ascension, X, and the declination, Y, for precession from epoch 1, DA (days), MA (months), YA (years), to epoch 2, DB (days), MB (months), YB (years). The coordinates X,Y are input at epoch 1 and output via P,Q at epoch 2. All angles in *radians*.

Set FL(10) to 0 on first call or whenever either or both of the epochs is changed. FL(10) is set to 1 by the routine, with a saving of execution time in subsequent calls for the same pair of epochs.

Other routine called: JULDAY (1100).

PRCESS2

It is sometimes necessary to make rather more exact calculations for precession than is done by PRCESS1. This is especially so if you wish to make large extrapolations into the future or the past, or if you are working in regions close to the north and south poles where the previous formulae do not work well. The routine PRCESS2 makes a rigorous reduction of equatorial coordinates from one epoch to another using matrices. The calculation proceeds in two parts.

2500 PRCESS1 and 2600 PRCESS2

First, the routine converts the given coordinates, correct at epoch 1 and input as X and Y as before, to their values for the epoch 2000 January 1.5 (J2000.0). Second, the routine converts these intermediate coordinates to the required values correct at epoch 2, and output via P and Q.

All this is transparent to you, the user. You can call PRCESS2 in exactly the same fashion as PRCESS1, setting the epochs via DA, MA, YA, and DB, MB, YB as before. However, you need to declare arrays MT(3,3), CV(3), HL(3), and MV(3,3) in the handling program and the time-saving flag is FL(11) rather than FL(10). You must also change the calling point from GOSUB 2500 to GOSUB 2600.

Formulae for PRCESS2

$XA = (0.6406161 \times T) + (8.39E\text{–}5 \times T^2) + (5.0E\text{–}6 \times T^3)$
$ZA = (0.6406161 \times T) + (3.041E\text{–}4 \times T^2) + (5.1E\text{–}6 \times T^3)$
$TA = (0.5567530 \times T) + (1.185E\text{–}4 \times T^2) - (1.16E\text{–}5 \times T^3)$
T = number of Julian centuries of 36525 days since J2000.0

These constants are calculated for each of the epochs, and are used to construct the transformation matrices.

Details of PRCESS2

Called by GOSUB 2600.

Corrects the right ascension, X, and the declination, Y, for precession from epoch 1, DA (days), MA (months), YA (years), to epoch 2, DB (Days), MB (months), YB (years). The coordinates X,Y are input at epoch 1 and output via P,Q at epoch 2. All angles in *radians*.

Set FL(11) to 0 on first call or whenever either or both of the epochs is changed. FL(11) is set to 1 by the routine, with a saving of execution time in subsequent calls for the same pair of epochs.

Declare the arrays MT(3,3), CV(3), HL(3), and MV(3,3) before calling the routine.

Other routine called: JULDAY (1100).

```
2497  REM
2498  REM      Subroutine PRCESS1 - approximate method
2499  REM
```

Line	Note	Code
2500	First call for these epochs; if not, don't repeat	`2500  IF FL(10)=1 THEN GOTO 2565`
2505	Epoch 1 . . .	`2505  FL(1)=0 : DY=DA : MN=MA : YR=YA`
2510	. . . use JULDAY to convert to Julian date, DK	`2510  GOSUB 1100 : DK=DJ : FL(10)=1`
2515	ER(1)=1 signifies impossible date	`2515  IF ER(1)=1 THEN FL(10)=0 : RETURN`
2520	Epoch 2 . . .	`2520  FL(1)=0 : DY=DB : MN=MB : YR=YB`
2525	. . . use JULDAY to convert to Julian date, DJ	`2525  GOSUB 1100 : T1=DK/36525 : T2=DJ/36525.0`
2530	T1 and T2 are centuries since 1900 January 0.5	`2530  IF ER(1)=1 THEN FL(10)=0 : RETURN`
2535	Find number of years between the two epochs, NY	`2535  NY=(DJ-DK)/365.2425`
2540	Calculate precession constants, M and N, . . .	`2540  M1=3.07234+(1.86E-3*T1) : TP=6.283185308`
		`2545  N1=20.0468-(8.5E-3*T1)`
2550	. . . for each epoch . . .	`2550  M2=3.07234+(1.86E-3*T2)`
		`2555  N2=20.0468-(8.5E-3*T2)`
2560	. . . and take the average to find MP, NP	`2560  MP=(M1+M2)/2.0 : NP=(N1+N2)/2.0`
2565	Now find change in X . . .	`2565  X1=(MP+(NP*SIN(X)*TAN(Y)/15.0))*7.272205E-5*NY`
2570	. . . and in Y and add them to X,Y to get results P,Q	`2570  Y1=NP*COS(X)*4.848137E-6*NY : P=X+X1 : Q=Y+Y1`
2575	Make sure that P is still in range 0–2π	`2575  IF P>TP THEN P=P-TP : GOTO 2575`
2580	Normal return from PRCESS1	`2580  IF P<0 THEN P=P+TP : GOTO 2580`
		`2585  RETURN`

Notes		Code
| | | ```
1    REM
2    REM        Handling program HPRCESS1
3    REM
``` |
| | | ```
5    DIM FL(20),ER(20),SW(20)
``` |
10	FNI needed by JULDAY	`10   DEF FNI(W)=SGN(W)*INT(ABS(W))`
15	Converts degrees to radians	`15   DEF FNM(W)=1.745329252E-2*W`
20	...and radians to degrees	`20   DEF FND(W)=5.729577951E1*W`
| | | ```
25   PRINT : PRINT
30   PRINT "Approximate precession in RA and DEC"
35   PRINT "----------------------------------"
40   PRINT
``` |
		`45   PRINT`
50	FL(10)=1 if the epochs have not changed	`50   IF FL(10)=1 THEN GOTO 75`
		`55   Q$="Precess from (D,M,Y) .............. "`
60	Get new epoch 1 ...	`60   PRINT Q$; : INPUT DA,MA,YA`
		`65   Q$=".........to (D,M,Y) .............. "`
70	...and epoch 2	`70   PRINT Q$; : INPUT DB,MB,YB : PRINT`
75	Now get the equatorial coordinates at epoch 1 ...	`75   Q$="Right ascension at epoch 1 (H,M,S) .. "`
		`80   PRINT Q$; : INPUT XD,XM,XS`
85	...in radians via MINSEC and FNM	`85   SW(1)=-1 : GOSUB 1000 : XA=FNM(X*15.0)`
		`90   Q$="Declination at epoch 1 (D,M,S) ...... "`
		`95   PRINT Q$; : INPUT XD,XM,XS`
		`100  SW(1)=-1 : GOSUB 1000 : Y=FNM(X) : X=XA`
105	Call PRCESS1, and convert right ascension at epoch 2 to degrees	`105  PRINT : GOSUB 2500 : X=FND(P/15.0)`
110	Was it an impossible date?	`110  IF ER(1)=1 THEN GOTO 145`
115	Convert to H,M,S form with MINSEC...	`115  SW(1)=1 : NC=9 : GOSUB 1000`
		`120  Q$="Right ascension at epoch 2 (H,M,S) .. "`
125	...and display new right ascension neatly	`125  PRINT Q$+"    "+MID$(OP$,4,12)`
130	Declination to D,M,S ...	`130  X=FND(Q) : GOSUB 1000`
		`135  Q$="Declination at epoch 2 (D,M,S) ...... "`
140	...and display it	`140  PRINT Q$+OP$`
145	Another go?	`145  Q$="Again (Y or N) .................... "`
150	Call YESNO to see	`150  PRINT : GOSUB 960`
		`155  IF E=0 THEN STOP`
160	If FL(10)=0 then we have not yet done anything	`160  IF FL(10)=0 THEN GOTO 45`
170	Set FL(10) back to 0 for new epochs	`165  Q$="Same epochs (Y or N) .............. "`
		`170  GOSUB 960 : FL(10)=E : GOTO 45`
| | | ```

INCLUDE YESNO, MINSEC, JULDAY, PRCESS1
``` |

Example

```
Approximate precession in RA and DEC
------------------------------------


Precess from (D,M,Y) ................. ? 0.9,1,1950
.........to (D,M,Y) ................. ? 4,2,1990

Right ascension at epoch 1 (H,M,S) .. ? 12,12,12
Declination at epoch 1 (D,M,S) ...... ? 23,23,23

Right ascension at epoch 2 (H,M,S) ..    12 14 14.00
Declination at epoch 2 (D,M,S) ......  + 23 10  0.62

Again (Y or N) ...................... ? Y
Same epochs (Y or N) ................ ? N

Precess from (D,M,Y) ................. ? 4,2,1990
.........to (D,M,Y) ................. ? 0.9,1,1950

Right ascension at epoch 1 (H,M,S) .. ? 12,14,14
Declination at epoch 1 (D,M,S) ...... ? 23,10,0.62

Right ascension at epoch 2 (H,M,S) ..    12 12 12.19
Declination at epoch 2 (D,M,S) ......  + 23 23 22.59

Again (Y or N) ...................... ? Y
Same epochs (Y or N) ................ ? N

Precess from (D,M,Y) ................. ? 0.9,1,1950
.........to (D,M,Y) ................. ? 1,1,2050

Right ascension at epoch 1 (H,M,S) .. ? 12,12,12
Declination at epoch 1 (D,M,S) ...... ? 23,23,23

Right ascension at epoch 2 (H,M,S) ..    12 17 16.35
Declination at epoch 2 (D,M,S) ......  + 22 50  1.96

Again (Y or N) ...................... ? Y
Same epochs (Y or N) ................ ? N

Precess from (D,M,Y) ................. ? 1,1,2050
.........to (D,M,Y) ................. ? 0.9,1,1950

Right ascension at epoch 1 (H,M,S) .. ? 12,17,16.35
Declination at epoch 1 (D,M,S) ...... ? 22,50,1.96

Right ascension at epoch 2 (H,M,S) ..    12 12 13.16
Declination at epoch 2 (D,M,S) ......  + 23 23 20.15

Again (Y or N) ...................... ? N
```

```
2597 REM
2598 REM        Rigorous precession in RA and DEC
2599 REM
```

2600	First call for these epochs; if not, don't repeat

```
2600 IF FL(11)=1 THEN GOTO 2670
```

2605	Epoch 1; call local routine to get constants...
2610	Was date impossible?
2615	...and form the matrix MT(3,3)

```
2605 DY=DA : MN=MA : YR=YA : GOSUB 2750

2610 IF ER(1)=1 THEN FL(11)=0 : RETURN
2615 MT(1,1)=C1*C3*C2-S1*S2 : MT(1,2)=-S1*C3*C2-C1*S2
2620 MT(1,3)=-S3*C2 : MT(2,1)=C1*C3*S2+S1*C2
2625 MT(2,2)=-S1*C3*S2+C1*C2 : MT(2,3)=-S3*S2
2630 MT(3,1)=C1*S3 : MT(3,2)=-S1*S3 : MT(3,3)=C3
```

2635	Epoch 2; call local routine to get constants...
2640	Was it impossible?
2645	...and form matrix MV(3,3)

```
2635 DY=DB : MN=MB : YR=YB : GOSUB 2750

2640 IF ER(1)=1 THEN FL(11)=0 : RETURN
2645 MV(1,1)=C1*C3*C2-S1*S2 : MV(1,2)=-S1*C3*C2-C1*S2
2650 MV(3,1)=-S3*C2 : MV(1,2)=C1*C3*S2+S1*C2
2655 MV(2,2)=-S1*C3*S2+C1*C2 : MV(3,2)=-S3*S2
2660 MV(1,3)=C1*S3 : MV(2,3)=-S1*S3 : MV(3,3)=C3
2665 FL(11)=1 : PI=3.141592654 : TP=2.0*PI

2669 REM Convert input into column vector
```

2670	Form the column vector CV(3) from input X,Y

```
2670 CN=COS(Y) : CV(1)=COS(X)*CN
2675 CV(2)=SIN(X)*CN : CV(3)=SIN(Y)

2679 REM Multiply MT by CV
```

2680	Multiply CV by MT to form new column vector...
2685	...correct for epoch 1.5,1,2000...
2695	...and put into CV

```
2680 FOR J=1 TO 3 : SM=0.0

2685 FOR I=1 TO 3 : SM=SM+MT(I,J)*CV(I) : NEXT I
2690 HL(J)=SM : NEXT J
2695 FOR I=1 TO 3 : CV(I)=HL(I) : NEXT I

2699 REM Multiply MV by CV
```

2700	Multiply CV by MV to form new column vector...
2705	...correct for epoch 2...
2715	...and put into CV

```
2700 FOR J=1 TO 3 : SM=0.0

2705 FOR I=1 TO 3 : SM=SM+MV(I,J)*CV(I) : NEXT I
2710 HL(J)=SM : NEXT J
2715 FOR I=1 TO 3 : CV(I)=HL(I) : NEXT I

2719 REM Convert column vector into output
```

2720	Convert CV into the output coordinates P,Q
2725	FNS returns inverse sine
2730	Correct for ambiguity of inverse tangent

```
2720 IF ABS(CV(1))<1E-20 THEN CV(1)=1E-20

2725 P=ATN(CV(2)/CV(1)) : Q=FNS(CV(3))

2730 IF CV(1)<0 THEN P=P+PI
```

2735	Make sure that P is in range $0-2\pi$

```
2735 IF P<0 THEN P=P+TP : GOTO 2735
2740 IF P>TP THEN P=P-TP : GOTO 2740
```

2745	Normal return from PRCESS2

```
2745 RETURN
```

2750	Local routine to finds precessional constants; call JULDAY
2755	T is Julian centuries since 2000 January 1.5
2760	Calculate XA, ZA, and TA...

```
2750 FL(1)=0 : GOSUB 1100 : T=(DJ-36525.0)/36525.0

2755 IF ER(1)=1 THEN RETURN

2760 XA=(((0.000005*T)+0.0000839)*T+0.6406161)*T
2765 ZA=(((0.0000051*T)+0.0003041)*T+0.6406161)*T
2770 TA=(((-0.0000116*T)-0.0001185)*T+0.556753)*T
```

2775	...but we really need their sines and cosines

```
2775 XA=FNM(XA) : ZA=FNM(ZA) : TA=FNM(TA)
2780 C1=COS(XA) : C2=COS(ZA) : C3=COS(TA)
2785 S1=SIN(XA) : S2=SIN(ZA) : S3=SIN(TA)
2790 RETURN
```

| | ```
1    REM
2    REM          Handling program HPRCESS2
3    REM

5    DIM FL(20),ER(20),SW(20)
10   DIM MT(3,3),CV(3),HL(3),MV(3,3)
``` |

Notes	Code
10 Declare the matrices needed by PRCESS2	
15 FNI needed by JULDAY	
20 Converts degrees to radians	
25 ... and radians to degrees	
30 FNS returns inverse sine	

```
15   DEF FNI(W)=SGN(W)*INT(ABS(W))
20   DEF FNM(W)=1.745329252E-2*W
25   DEF FND(W)=5.729577951E1*W
30   DEF FNS(W)=ATN(W/(SQR(1-W*W)+1E-20))

35   PRINT : PRINT
40   PRINT "Rigorous precession in RA and DEC"
45   PRINT "--------------------------------" : PRINT

50   PRINT
55   IF FL(11)=1 THEN GOTO 80
60   Q$="Precess from (D,M,Y) ................ "
65   PRINT Q$; : INPUT DA,MA,YA
70   Q$=".........to (D,M,Y) ................ "
75   PRINT Q$; : INPUT DB,MB,YB : PRINT

80   Q$="Right ascension at epoch 1 (H,M,S) .. "
85   PRINT Q$; : INPUT XD,XM,XS
90   SW(1)=-1 : GOSUB 1000 : XA=FNM(X*15.0)
95   Q$="Declination at epoch 1 (D,M,S) ...... "
100  PRINT Q$; : INPUT XD,XM,XS
105  SW(1)=-1 : GOSUB 1000 : Y=FNM(X) : X=XA

110  PRINT : GOSUB 2600 : X=FND(P/15.0)

115  IF ER(1)=1 THEN GOTO 150

120  SW(1)=1 : NC=9 : GOSUB 1000
125  Q$="Right ascension at epoch 2 (H,M,S) .. "

130  PRINT Q$+"    "+MID$(OP$,4,12)

135  X=FND(Q) : GOSUB 1000
140  Q$="Declination at epoch 2 (D,M,S) ...... "
145  PRINT Q$+OP$

150  Q$="Again (Y or N) .................... "
155  PRINT : GOSUB 960
160  IF E=0 THEN STOP
165  IF FL(11)=0 THEN GOTO 50
170  Q$="Same epochs (Y or N) ................ "
175  GOSUB 960
180  FL(11)=E : GOTO 50

INCLUDE YESNO, MINSEC, JULDAY, PRCESS2
```

Notes (left column):

55   FL(11)=1 if the epochs have not changed

65   Get new epoch 1 ...

75   ... and epoch 2

80   Now get the equatorial coordinates at epoch 1 ...

90   ... in radians via MINSEC and FNM

110   Call PRCESS2, and convert right ascension at epoch 2 to degrees

115   Was there an impossible date?

120   Convert to H,M,S form with MINSEC ...

130   ... and display new right ascension neatly

135   Declination to D,M,S ...

145   ... and display it

150   Another go?

155   Call YESNO to see

165   If FL(11)=0 then we have not yet done anything

180   Set FL(11) back to 0 for new epochs
```

2500 PRCESS1 and 2600 PRCESS2

Example

```
Rigorous precession in RA and DEC
---------------------------------


Precess from (D,M,Y) ................. ? 0.9,1,1950
.........to (D,M,Y) ................. ? 4,2,1990

Right ascension at epoch 1 (H,M,S) .. ? 12,12,12
Declination at epoch 1 (D,M,S) ...... ? 23,23,23

Right ascension at epoch 2 (H,M,S) ..    12 14 13.93
Declination at epoch 2 (D,M,S) ......  + 23 10  0.64

Again (Y or N) ...................... ? Y
Same epochs (Y or N) ................ ? N

Precess from (D,M,Y) ................. ? 4,2,1990
.........to (D,M,Y) ................. ? 0.9,1,1950

Right ascension at epoch 1 (H,M,S) .. ? 12,14,13.93
Declination at epoch 1 (D,M,S) ...... ? 23,10,0.64

Right ascension at epoch 2 (H,M,S) ..    12 12 12.00
Declination at epoch 2 (D,M,S) ......  + 23 23 23.00

Again (Y or N) ...................... ? Y
Same epochs (Y or N) ................ ? N

Precess from (D,M,Y) ................. ? 0.9,1,1950
.........to (D,M,Y) ................. ? 1,1,2050

Right ascension at epoch 1 (H,M,S) .. ? 12,12,12
Declination at epoch 1 (D,M,S) ...... ? 23,23,23

Right ascension at epoch 2 (H,M,S) ..    12 17 15.84
Declination at epoch 2 (D,M,S) ......  + 22 50  2.86

Again (Y or N) ...................... ? Y
Same epochs (Y or N) ................ ? N

Precess from (D,M,Y) ................. ? 1,1,2050
.........to (D,M,Y) ................. ? 0.9,1,1950

Right ascension at epoch 1 (H,M,S) .. ? 12,17,15.84
Declination at epoch 1 (D,M,S) ...... ? 22,50,2.86

Right ascension at epoch 2 (H,M,S) ..    12 12 12.00
Declination at epoch 2 (D,M,S) ......  + 23 23 23.00

Again (Y or N) ...................... ? N
```

2800 PARALLX

This routine converts the geocentric hour angle and declination to the apparent hour angle and declination, and vice-versa, making allowance for geocentric parallax.

The equatorial coordinates of members of the Solar System calculated in later subroutines are those appropriate to an observer situated at the centre of the Earth (*geocentric* coordinates). When the celestial body in question is at a very large distance away, the same coordinates apply to observations made from the Earth's surface. However, objects within the Solar System are sufficiently close that their apparent positions (*topocentric* coordinates) change slightly depending on the exact vantage point. The Sun, for example, may appear displaced by as much as 8 arcseconds from its geocentric position, and the Moon by as much as 1 degree.

Routine PARALLX calculates the apparent hour angle (P) and declination (Q) for an observer at a given geographical latitude (GP) and height above sea-level (HT), when the geocentric hour angle (X), declination (Y), and equatorial horizontal parallax (HP) are known. It will also perform the reverse operation, calculating the geocentric coordinates from the apparent position. The correction for parallax to the geocentric hour angle (DX) is the same as that to the geocentric right ascension. Having run the routine, the apparent right ascension, RA1, may he found from

RA1 = RA − DX,

where RA is the geocentric right ascension.

Execution of PARALLX is controlled by the flag FL(12). This should be set to 0 before calling the routine for the first time, or whenever new values of GP or HT are specified. It is set to 1 in the routine (line 2815), signalling that the calculations applying only to the geographical location have already been made, and need not be repeated. Subsequent calls with FL(12) = 1 skip lines 2805–2825 inclusive, thus saving execution time.

The direction of conversion is indicated by the switch SW(5). When the switch has the value +1, then X,Y are treated as geocentric coordinates and P,Q are the corresponding apparent coordinates. When SW(5) = −1, the conversion is from apparent coordinates (X,Y) to geocentric coordinates (P,Q). Within the routine itself, the conversion from true to apparent coordinates is

2800 PARALLX

performed by lines 2870–2900. There is no corresponding section for conversion the other way. Instead, PARALLX adopts an iterative procedure (lines 2845–2860), treating the current position as the true position, calculating the corresponding apparent position, comparing it with the input apparent position, and making adjustments until agreement is reached within 10^{-6} radians (line 2855).

The intrinsic parallax of the celestial body is indicated by HP, the equatorial horizontal parallax. This is the angle subtended at the body by the Earth's equatorial radius. If RP is the distance of the body from the centre of the Earth in kilometres, then

$$SIN(HP) = 6378.16/RP.$$

HP is calculated explicitly for the Moon by subroutine MOON (6000). Its value for the Sun is given approximately by

$$HP = 8.794/RR \text{ arcseconds,}$$

where RR is the Sun–Earth distance in AU. (One AU is the semi-major axis of the Earth's orbit = 1.496×10^8 kilometres.)

Formulae

$$P = X + DX$$
$$TAN(Q) = COS(P) \times ((RP \times SIN(Y)) - RS)/((RP \times COS(Y) \times COS(X)) - RC)$$
$$TAN(DX) = RC \times SIN(X)/(RP \times COS(Y)) - (RC \times COS(Y)))$$
$$RP = 1/SIN(HP)$$
$$RC = COS(U) + (HT \times COS(GP))$$
$$RS = (9.96647E{-}1 \times SIN(U)) + (HT \times SIN(GP))$$
$$TAN(U) = 9.96647E{-}1 \times SIN(GP)/COS(GP)$$

Details of PARALLX

Called by GOSUB 2800.

Converts X, the true or apparent hour angle, and Y, the true or apparent declination, into P, the apparent or true hour angle, and Q, the apparent or true declination, for a given GP, the geographical latitude, HT, the height above sea-level, and HP, the equatorial horizontal parallax. All angles in *radians*. HT is expressed as a fraction of the Earth's radius of 6378.16 kilometres.

Set FL(12) to 0 on first call or for new values of GP or HT. The flag is set to 1 by the routine. Subsequent calls with FL(12) = 1 execute slightly faster by skipping calculations involving GP and HT which do not need to be repeated.

Set SW(5) = +1 for true to apparent conversion.

Set SW(5) = −1 for apparent to true conversion.

		2797	REM
		2798	REM Subroutine PARALLX
		2799	REM
2800	First call for this location?	2800	IF FL(12)=1 THEN GOTO 2830
2805	Calculate trigonometric quantities once only	2805	C1=COS(GP) : S1=SIN(GP)
		2810	U=ATN(9.96647E-1*S1/C1)
		2815	C2=COS(U) : S2=SIN(U) : FL(12)=1
		2820	RS=(9.96647E-1*S2)+(HT*S1)
2825	TP is 2π	2825	RC=C2+(HT*C1) : TP=6.283185308
2830	RP is distance of object in Earth radii	2830	RP=1/SIN(HP)
2835	If true to apparent call local routine once; return	2835	IF SW(5)=1 THEN GOSUB 2870 : RETURN
2840	Apparent to true; set up iterative loop	2840	X1=X : Y1=Y : P1=0 : Q1=0
2845	Call local parallax routine	2845	GOSUB 2870 : P2=P-X : Q2=Q-Y
		2850	A=ABS(P2-P1) : B=ABS(Q2-Q1)
2855	Close enough to the solution? . . .	2855	IF A<1E-6 AND B<1E-6 THEN GOTO 2865
2860	. . . no; once more round the loop	2860	X=X1-P2 : Y=Y1-Q2 : P1=P2 : Q1=Q2 : GOTO 2845
2865	. . . yes; calculate corrected coordinates P,Q and return	2865	P=X1-P2 : Q=Y1-Q2 : X=X1 : Y=Y1 : RETURN
2870	Local parallax routine; find trigonometric values once only	2870	CX=COS(X) : SY=SIN(Y) : CY=COS(Y)
		2875	A=(RC*SIN(X))/((RP*CY)-(RC*CX))
2880	DX is the correction to the hour angle	2880	DX=ATN(A) : P=X+DX : CP=COS(P)
2885	Make sure P is in range $0-2\pi$	2885	IF P>TP THEN P=P-TP : GOTO 2885
		2890	IF P<0 THEN P=P+TP : GOTO 2890
2895	Find Q	2895	Q=ATN(CP*(RP*SY-RS)/(RP*CY*CX-RC))
2900	Return from local parallax routine	2900	RETURN

Notes		Code
	1	`REM`
	2	`REM        Handling program HPARALLX`
	3	`REM`
10 FNI is needed by JULDAY	5	`DIM FL(20),ER(20),SW(20)`
15 Converts degrees to radians ...	10	`DEF FNI(W)=SGN(W)*INT(ABS(W))`
20 ... and radians to degrees	15	`DEF FNM(W)=1.745329252E-2*W`
	20	`DEF FND(W)=5.729577951E1*W`
	25	`PRINT : PRINT : SH=0`
	30	`PRINT "Corrections for parallax in HA and DEC"`
	35	`PRINT "-------------------------------------"`
	40	`PRINT : PRINT`
	45	`PRINT`
50 FL(12)=1 means that the location is same as before	50	`IF FL(12)=1 THEN GOTO 110`
	55	`Q$="Geographical longitude (W neg; D,M,S)  "`
60 Get a new location ...	60	`PRINT Q$; : INPUT XD,XM,XS`
65 ... longitude in degrees (GL) via MINSEC	65	`SW(1)=-1 : GOSUB 1000 : GL=X`
	70	`Q$="Geographical latitude  (S neg; D,M,S)  "`
75 ... latitude in radians (GP) via MINSEC and FNM	75	`PRINT Q$; : INPUT XD,XM,XS : GOSUB 1000 : GP=FNM(X)`
	80	`Q$="Height above sea-level (metres) ...... "`
85 ... height above sea-level (HT) in Earth radii	85	`PRINT Q$; : INPUT X : HT=X/6378140.0`
	90	`Q$="Daylight saving (H ahead of zone t) .. "`
95 ... the time bits: DS for hours ahead of zone time	95	`PRINT Q$; : INPUT DS`
	100	`Q$="Time zone (hours; West negative) ..... "`
105 ... TZ for hours ahead of UT	105	`PRINT Q$; : INPUT TZ`
110 FL(3)=1 means same date as before	110	`IF FL(3)=1 THEN GOTO 125`
	115	`Q$="Calendar date (D,M,Y) .............. "`
120 Get a new date	120	`PRINT : PRINT Q$; : INPUT DY,MN,YR`
125 SH is a local switch meaning 'same time' if set to 1	125	`IF SH=1 THEN GOTO 145`
	130	`Q$="Local civil time (H,M,S) ............ "`
135 Get a new local civil time ...	135	`PRINT Q$; : INPUT XD,XM,XS`
140 ... in decimal hours (via MINSEC) in TM	140	`SW(1)=-1 : GOSUB 1000 : TM=X`
145 Call TIME to find local sidereal time	145	`SW(2)=1 : GOSUB 1300`
150 Impossible date? (JULDAY)	150	`IF ER(1)=1 THEN GOTO 310`
155 Don't tell us all this more than once	155	`IF SH=1 THEN GOTO 190`
160 Display all the times found in TIME	160	`Q$="Universal time (H,M,S) .............. "`
	165	`PRINT Q$+UT$`
	170	`Q$="Greenwich sidereal time (H,M,S) ...... "`
	175	`PRINT Q$+SG$`
	180	`Q$="Local sidereal time (H,M,S) ......... "`
	185	`PRINT Q$+TL$`
190 Which direction to convert?	190	`Q$="True to apparent (Y or N) ........... "`
195 Call YESNO for a response	195	`PRINT : GOSUB 960 : SW(5)=1`
	200	`IF E=1 THEN GOTO 220`
	205	`Q$="Apparent to true (Y or N) ........... "`
	210	`GOSUB 960 : SW(5)=-1`
215 Skip to the end if neither direction	215	`IF E=0 THEN GOTO 310`
220 Get the uncorrected coordinates ...	220	`Q$="Right-ascension (H,M,S) .............. "`
	225	`PRINT : PRINT Q$; :  INPUT XD,XM,XS`

230	. . . right ascension in decimal hours; call HRANG	230	`SW(1)=-1 : GOSUB 1000 : GOSUB 1600`
235	. . . convert hour angle to H,M,S form . . .	235	`X=P : SW(1)=1 : NC=9 : GOSUB 1000 : SW(1)=-1`
		240	`Q$="Hour angle (H,M,S) . . . . . . . . . . . . . . . . . . "`
245	. . . and tell us what it is	245	`PRINT Q$+" "+MID$(OP$,4,12)`
		250	`Q$="Declination (D,M,S) . . . . . . . . . . . . . . . "`
255	. . . declination in radians (MINSEC and FNM)	255	`PRINT Q$; : INPUT XD,XM,XS : GOSUB 1000 : Y=FNM(X)`
		260	`Q$="Equatorial horizontal parallax (D,M,S) "`
265	. . . horizontal parallax . . .	265	`PRINT Q$; : INPUT XD,XM,XS : GOSUB 1000`
270	. . . and hour angle in radians; call PARALLX	270	`HP=FNM(X) : X=FNM(P*15.0) : GOSUB 2800`
275	Corrected hour angle to hours; call HRANG	275	`X=FND(P/15.0) : GOSUB 1600 : PRINT`
280	Display the results . . .	280	`Q$="Corrected right-ascension (H,M,S) . . . . "`
285	. . . in H,M,S form . . .	285	`X=P : SW(1)=1 : NC=9 : GOSUB 1000`
290	. . . neatly	290	`PRINT Q$+"       "+MID$(OP$,4,12)`
		295	`Q$="Corrected declination (D,M,S) . . . . . . . . "`
		300	`X=FND(Q) : GOSUB 1000`
		305	`PRINT Q$+OP$`
310	Another go?	310	`Q$="Again (Y or N) . . . . . . . . . . . . . . . . . . . . . . "`
315	YESNO gets the answer	315	`PRINT : GOSUB 960`
		320	`IF E=0 THEN STOP`
325	If FL(12)=0 we have not yet done anything	325	`IF FL(12)=0 THEN GOTO 45`
		330	`Q$="Same place (Y or N) . . . . . . . . . . . . . . . . . "`
335	FL(12)=0 signifies a new place	335	`GOSUB 960 : FL(12)=E`
		340	`Q$="Same date (Y or N) . . . . . . . . . . . . . . . . . . "`
345	FL(3)=0 signifies a new date	345	`GOSUB 960 : FL(3)=E`
		350	`Q$="Same time (Y or N) . . . . . . . . . . . . . . . . . . "`
355	SH=0 signifies a new time	355	`GOSUB 960 : SH=E : GOTO 45`

`INCLUDE YESNO, MINSEC, JULDAY, TIME, HRANG, PARALLX`

2800 PARALLX

Example

```
Corrections for parallax in HA and DEC
---------------------------------------


Geographical longitude (W neg; D,M,S)  ? -100,0,0
Geographical latitude  (S neg; D,M,S)  ? 50,0,0
Height above sea-level (metres) ......  ? 60
Daylight saving (H ahead of zone t) ..  ? 0
Time zone (hours; West negative) .....  ? -6

Calendar date (D,M,Y) ................  ? 26,2,1979
Local civil time (H,M,S) .............  ? 10,45,0
Universal time (H,M,S) ...............    16 45  0.00
Greenwich sidereal time (H,M,S) ......     3  8 44.80
Local sidereal time (H,M,S) ..........    20 28 44.80

True to apparent (Y or N) ............  ? Y

Right ascension (H,M,S) ..............  ? 22,35,19
Hour angle (H,M,S) ...................    21 53 25.80
Declination (D,M,S) ..................  ? -7,41,13
Equatorial horizontal parallax (D,M,S) ? 1,1,9

Corrected right ascension (H,M,S) ....    22 36 43.22
Corrected declination (D,M,S) ........  -  8 32 17.40

Again (Y or N) .......................  ? Y
Same place (Y or N) ..................  ? Y
Same date (Y or N) ...................  ? Y
Same time (Y or N) ...................  ? Y


True to apparent (Y or N) ............  ? N
Apparent to true (Y or N) ............  ? Y

Right ascension (H,M,S) ..............  ? 22,36,43.22
Hour angle (H,M,S) ...................    21 52  1.58
Declination (D,M,S) ..................  ? -8,32,17.40
Equatorial horizontal parallax (D,M,S) ? 1,1,9

Corrected right ascension (H,M,S) ....    22 35 19.00
Corrected declination (D,M,S) ........  -  7 41 13.00

Again (Y or N) .......................  ? N
```

3000 REFRACT

This routine converts the true altitude into the apparent altitude, and vice-versa, correcting for atmospheric refraction.

Observations from the Earth's surface have to be made through the atmosphere. Although the air is tenuous, it has a refractive index which is significantly greater than 1, the value for a vacuum. When we observe a celestial object at an angle from the vertical the rays are bent slightly away from the path they would have followed if the atmosphere had not been present. The atmosphere clings to the curve of the Earth and it acts rather like a weak lens, making stars appear to be nearer to the vertical than they really are. The largest effect is seen at rising and setting where the horizontal refraction is about 34 minutes of arc. Since the effect is always to increase the Sun's apparent altitude, the length of the day is increased because of it.

Subroutine REFRACT assumes a model atmosphere whose refractive properties depend only on pressure, PR, and temperature, TR, at the place of observation. The amount of refraction actually observed depends on the detailed disposition of the atmosphere along the line of sight to the celestial object in question, and significant deviations from the values calculated here must be expected in some instances, especially near the horizon. Nevertheless, REFRACT will provide a very good approximation which will be more than adequate in most circumstances.

The altitude of the object is input via the parameter Y, corrected by the routine, and returned via the parameter Q. The direction of correction is indicated by the switch SW(6). When SW(6) = −1, the true value of the altitude is input and corrected to give the apparent altitude. When SW(6) = +1, the apparent altitude is input and transformed into the true altitude. REFRACT works for all values of altitude above about −5 degrees (5 degrees below the horizon).

The transformation of the apparent altitude into the true altitude is straightforward and involves a single pass through the calculation. The calculation of the refraction, RF, due to the atmosphere (lines 3050–3070) uses the apparent altitude as an argument. Converting a true altitude into its apparent value is therefore not so easy since the conversion process requires the answer in order

3000 REFRACT

to calculate the answer. REFRACT overcomes this problem with an iterative procedure (lines 3020–3030). It assumes that the current value of Y *is* the apparent value and calculates the corresponding true value. If this does not agree with the input value, it corrects the current apparent value and tries again until agreement is reached.

Formulae

$Y_t = Y_a + RF$

Y_t = true altitude

Y_a = apparent altitude

If Y is less than 15 degrees:

$RF = A/B$

$A = PR \times (1.594E{-}1 + (1.96E{-}2 \times Y_a) + (2E{-}5 \times Y_a^2))$

$B = (273 + TR) \times (1 + (5.05E{-}1 \times Y_a) + (8.45E{-}2 \times Y_a^2))$

(Y,RF in degrees)

If Y is greater than 15 degrees:

$RF = 7.888888E{-}5 \times PR/((273 + TR) \times TAN(Y_a))$

(RF in radians)

In both cases:

PR = atmospheric pressure in millibars

TR = atmospheric temperature in degrees Centigrade

Details of REFRACT

Called by GOSUB 3000.

Transforms Y, the true or apparent altitude (radians), into Q, the corresponding apparent or true altitude (radians), for a model atmosphere with pressure PR millibars and temperature TR degrees Centigrade.

Set $SW(6) = -1$ for transformation from real to apparent.

Set $SW(6) = +1$ for transformation from apparent to real.

```
2997 REM
2998 REM          Subroutine REFRACT
2999 REM
```

3000 Which direction? . . .

```
3000 IF SW(6)=-1 THEN GOTO 3015
```

3005 Apparent to true; find refraction and correct altitude

```
3005 GOSUB 3035 : Q=Y+RF
```

3010 Normal return from REFRACT

```
3010 RETURN
```

3015 True to apparent; set up interative loop . . .

```
3015 Y1=Y : Y2=Y : R1=0
```

3020 . . . find refraction from local routine . . .

```
3020 Y=Y1+R1 : Q=Y : GOSUB 3035 : R2=RF
```

3025 . . . are we there yet? Return from REFRACT if we are

```
3025 IF R2=0 OR ABS(R2-R1)<1E-6 THEN Y=Y2 : RETURN
```

3030 . . . no, not quite; go around the loop again

```
3030 R1=R2 : GOTO 3020
```

3035 Find refraction for altitudes above 15 degrees

```
3035 IF Y<2.617994E-1 THEN GOTO 3050
```

3040 SW(6) determines the sign of the refraction

```
3040 RF=-SW(6)*7.888888E-5*PR/((273+TR)*TAN(Y))
3045 RETURN
```

3050 Find refraction for altitudes below 15 degrees

```
3050 IF Y<-8.7E-2 THEN RF=0 : RETURN
```

3055 . . . no refraction if below −5 degrees

```
3055 YD=Y*5.729578E1
3060 A=((2E-5*YD+1.96E-2)*YD+1.594E-1)*PR
3065 B=(273+TR)*((8.45E-2*YD+5.05E-1)*YD+1.0)
```

3070 SW(6) determines the sign of the refraction

```
3070 RF=-(A/B)*1.745329E-2*SW(6) : RETURN
```

	1	REM
	2	REM Handling program HREFRACT
	3	REM
	5	DIM FL(20),ER(20),SW(20)
10 Needed by JULDAY	10	DEF FNI(W)=SGN(W)*INT(ABS(W))
15 Degrees to radians . . .	15	DEF FNM(W)=1.745329252E-2*W
20 . . . and radians to degrees	20	DEF FND(W)=5.729577951E1*W
25 FNS returns inverse sine	25	DEF FNS(W)=ATN(W/(SQR(1-W*W)+1E-20))
30 FNC returns inverse cosine	30	DEF FNC(W)=1.570796327-FNS(W)
	35	PRINT : PRINT : SH=0 : OA=0
	40	PRINT "Atmospheric refraction"
	45	PRINT "---------------------" : PRINT
50 FL(4)=1 signifies same place as before	50	IF FL(4)=1 THEN GOTO 105
55 Get all the location details . . .	55	Q$="Daylight saving (H ahead of zone t) "
60 . . . daylight saving DS in hours	60	PRINT : PRINT Q$; : INPUT DS
	65	Q$="Time zone (hours; West negative) ... "
70 . . . time zone TZ in hours	70	PRINT Q$; : INPUT TZ
	75	Q$="Geog. longitude (D,M,S; W neg.) .... "
80 . . . longitude GL in degrees via MINSEC	80	PRINT Q$; : INPUT XD,XM,XS
	85	SW(1)=-1 : GOSUB 1000 : GL=X
90 . . . latitude GP in radians	90	Q$="Geog. latitude (D,M,S; S neg.) ..... "
	95	PRINT Q$; : INPUT XD,XM,XS
	100	SW(1)=-1 : GOSUB 1000 : GP=FNM(X)
105 FL(3)=1 means date has not changed since last time	105	IF FL(3)=1 THEN PRINT : GOTO 120
	110	Q$="Calendar date (D,M,Y) .... "
115 Get a new calendar date	115	PRINT : PRINT Q$; : INPUT DY,MN,YR
120 SH is local switch meaning 'same time'	120	IF SH=1 THEN GOTO 140
	125	Q$="Local civil time (H,M,S) .... "
130 Get a new local civil time TM . . .	130	PRINT Q$; : INPUT XD,XM,XS
135 . . . in hours via MINSEC	135	SW(1)=-1 : GOSUB 1000 : TM=X
140 Call TIME to find local sidereal time	140	SW(2)=1 : GOSUB 1300
145 ER(1)=1 means that the date is impossible (JULDAY)	145	IF ER(1)=1 THEN GOTO 370
150 Tell us the time just once . . .	150	IF SH=1 THEN GOTO 185
	155	Q$="Universal time (H,M,S) .... "
	160	PRINT Q$+UT$
	165	Q$="Greenwich sidereal time (H,M,S) .... "
	170	PRINT Q$+SG$
	175	Q$="Local sidereal`time (H,M,S) .... "
	180	PRINT Q$+TL$
185 OA is a local switch meaning 'same atmosphere'	185	IF OA=1 THEN GOTO 210
	190	Q$="Atmospheric pressure (mBar) .... "
195 Get new values for the atmosphere	195	PRINT : PRINT Q$; : INPUT PR
	200	Q$="Atmospheric temperature (C) .... "
	205	PRINT Q$; : INPUT TR
210 Which direction to convert?	210	Q$="True to apparent (Y or N) .... "
215 YESNO will tell us	215	SW(6)=-1 : PRINT : GOSUB 960
	220	IF E=1 THEN GOTO 240
	225	Q$="Apparent to true (Y or N) .... "
	230	SW(6)=+1 : GOSUB 960
235 If neither direction, skip to the end	235	IF E=0 THEN GOTO 370

240	Get the uncorrected coordinates . . .	240 `Q$="Right ascension (H,M,S) ........... "`
250	. . . right ascension in hours; then call HRANG	245 `PRINT : PRINT Q$; : INPUT XD,XM,XS` 250 `SW(1)=-1 : GOSUB 1000 : GOSUB 1600`
255	. . . convert hour angle to H,M,S form . . .	255 `X=P : NC=9 : SW(1)=1 : GOSUB 1000` 260 `Q$="Hour angle (H,M,S) ............... "`
265	. . . and display it neatly	265 `PRINT Q$+"   "+MID$(OP$,4,12)` 270 `Q$="Declination (D,M,S) ............. "`
275	. . . declination . . .	275 `PRINT Q$; : INPUT XD,XM,XS`
280	. . . and hour angle both in radians	280 `SW(1)=-1 : GOSUB 1000 : Y=FNM(X) : X=FNM(P*15.0)`
285	Find the uncorrected altitude by calling EQHOR . . .	285 `GOSUB 1500 : X=FND(Q)` 290 `NC=9 : SW(1)=1 : GOSUB 1000`
295	. . . and display it in D,M,S format	295 `Q$="Uncorrected altitude (D,M,S) ....... "`
300	Now we call REFRACT . . .	300 `PRINT Q$+OP$ : Y=Q : GOSUB 3000`
305	. . . and display the altitude correction in D,M,S form	305 `X=FND(RF) : GOSUB 1000` 310 `Q$="Altitude correction (D,M,S) ........ "` 315 `PRINT : PRINT Q$+OP$`
320	Call EQHOR to convert back to hour angle and declination	320 `X=P : Y=Q : GOSUB 1500 : P=FND(P) : Q=FND(Q)`
325	Convert corrected hour angle to H,M,S form	325 `X=P/15.0 : NC=9 : SW(1)=1 : GOSUB 1000` 330 `Q$="Corrected hour angle (H,M,S) ....... "`
335	. . . and display it neatly	335 `PRINT Q$+"   "+MID$(OP$,4,12)`
340	Call HRANG to find the corrected right ascension . . .	340 `GOSUB 1600 : X=P : GOSUB 1000`
345	. . . and display it in H,M,S form . . .	345 `Q$="Corrected right ascension (H,M,S) .. "`
350	. . . neatly	350 `PRINT Q$+"   "+MID$(OP$,4,12)` 355 `X=Q : GOSUB 1000` 360 `Q$="Corrected declination (D,M,S) ...... "`
365	Display the corrected declination in D,M,S format	365 `PRINT Q$+OP$`
370	Another go?	370 `Q$="Again (Y or N) .................... "`
375	Call YESNO for the answer	375 `PRINT : GOSUB 960` 380 `IF E=0 THEN STOP`
385	If FL(4)=0 we have not yet done anything	385 `IF FL(4)=0 THEN GOTO 50` 390 `Q$="Same place (Y or N) ............... "`
395	FL(4)=0 for new place	395 `GOSUB 960 : FL(4)=E` 400 `Q$="Same date (Y or N) ................ "`
405	FL(3)=0 for new date	405 `GOSUB 960 : FL(3)=E` 410 `Q$="Same time (Y or N) ................. "`
415	SH=0 for new time	415 `GOSUB 960 : SH=E` 420 `Q$="Same atmosphere (Y or N) ........... "`
425	OA=0 for new atmospheric parameters	425 `GOSUB 960 : OA=E : GOTO 50`

`INCLUDE YESNO, MINSEC, JULDAY, TIME, EQHOR, HRANG, REFRACT`

3000 REFRACT

Example

```
Atmospheric refraction
----------------------


Daylight saving (H ahead of zone t)  ? 0
Time zone (hours; West negative) ...  ? 0
Geog. longitude (D,M,S; W neg.) ....  ? 0,10,12
Geog. latitude (D,M,S; S neg.) .....  ? 51,12,13

Calendar date (D,M,Y) ..............  ? 23,3,1987
Local civil time (H,M,S) ...........  ? 1,1,24
Universal time (H,M,S) .............    1  1 24.00
Greenwich sidereal time (H,M,S) ....   13  1 22.46
Local sidereal time (H,M,S) ........   13  2  3.26

Atmospheric pressure (mBar) ........  ? 1012
Atmospheric temperature (C) ........  ? 21.7

True to apparent (Y or N) ..........  ? Y

Right ascension (H,M,S) ............  ? 23,14,0
Hour angle (H,M,S) .................    13 48  3.26
Declination (D,M,S) ................  ? 40,10,0
Uncorrected altitude (D,M,S) .......  +  4 22  2.41

Altitude correction (D,M,S) ........  +  0 10 11.25
Corrected hour angle (H,M,S) .......    13 48 18.51
Corrected right ascension (H,M,S) ..    23 13 44.74
Corrected declination (D,M,S) ......  + 40 19 45.77

Again (Y or N) .....................  ? Y
Same place (Y or N) ................  ? Y
Same date (Y or N) .................  ? Y
Same time (Y or N) .................  ? Y
Same atmosphere (Y or N) ...........  ? Y


True to apparent (Y or N) ..........  ? N
Apparent to true (Y or N) ..........  ? Y

Right ascension (H,M,S) ............  ? 23,13,44.74
Hour angle (H,M,S) .................    13 48 18.52
Declination (D,M,S) ................  ? 40,19,45.77
Uncorrected altitude (D,M,S) .......  +  4 32 13.68

Altitude correction (D,M,S) ........  -  0 10 11.25
Corrected hour angle (H,M,S) .......    13 48  3.26
Corrected right ascension (H,M,S) ..    23 14  0.00
Corrected declination (D,M,S) ......  + 40 10  0.01

Again (Y or N) .....................  ? N
```

3100 RISET

This routine calculates the local sidereal times and azimuths of rising and setting

During the course of one day, the stars and other celestial objects appear to move in circles about a single point in the sky. This point is the north celestial pole for observers in the northern hemisphere or the south celestial pole in the southern hemisphere, and it marks the intersection of the Earth's spin axis with the celestial sphere. Stars which are close enough to the pole remain above the horizon all day and are called *circumpolar* stars. Stars which are far enough away from the pole spend part of the day below the horizon. As they cross the horizon on the way down they *set*, and as they cross on the way up they *rise*. Stars further still from the pole spend all day below the horizon and are never seen in that location.

Routine RISET finds the circumstances of rising and setting for a celestial object whose right ascension and declination are known. The local sidereal times are calculated for a given geographical latitude, GP, and are returned via the variables LU (rising) and LD (setting). The azimuths are also given by AU and AD. the local sidereal times may be converted to Greenwich sidereal times, Universal Times, and local civil times using routine TIME (1300).

The coordinates input to the routine by X and Y are usually the true geocentric coordinates. However, there are several effects including atmospheric refraction (REFRACT, 3000) and parallax (PARALLX, 2800) which cause a small shift in the apparent position of a celestial object. This results in a vertical displacement at the horizon such that the apparent position has a higher altitude than the true position, an effect which can make several minutes difference to the times of rising and setting. RISET takes account of this by accepting the input parameter DI, the vertical displacement from the true position at the horizon. DI is positive if the apparent position has a larger altitude than the true position (i.e. it is above the true position). Atmospheric refraction is usually assumed to produce a vertical shift of 34 minutes of arc at the horizon; DI would then take the value $+9.89 \times 10^{-3}$ (radians). On the other hand, if your horizon has hills on it such that your apparent horizon is, say, 1 degree above sea-level, you can take this into account by subtracting 1.75×10^{-2} (radians)

3100 RISET

from the atmospheric displacement. Generally, DI is positive if the shift causes the object to spend longer above the horizon.

Execution of the subroutine is controlled by the flag FL(4). This must be set to 0 when calling the routine for the first time, but is set to one in line 3110. Subsequent calls then save time by skipping lines 3105 and 3110 which involve calculations concerned only with the observer's position. If a new geographical latitude is specified, set FL(4) = 0 before calling the routine. Note that the same flag is used in the same way by other routines.

Error conditions which arise during the execution of RISET are indicated by the error flag ER(3). This would normally be set to 0 by the routine (line 3140) indicating that no error had been detected. However, if the star is such that it never rises, ER(3) assumes the value +1, while if it is circumpolar, it is set to −1. In all cases, new values of LU, LD, AU, and AD are *not* calculated unless ER(3) = 0.

Formulae

$LU = [24 + X - H]_{24}$

$LD = [X + H]_{24}$

$COS(AU) = (SIN(Y)+(SIN(DI) \times SIN(GP))/(COS(DI) \times COS(GP))$

$AD = 360° - AU$

$COS(H) = -(SIN(DI)+(SIN(GP) \times SIN(Y))/(COS(GP) \times COS(Y))$

$[\ ]_{24}$ indicates reduction to the range 0–24 by the addition or subtraction of multiples of 24.

Details of RISET

Called by GOSUB 3100.

Calculates the local sidereal times (hours), of rising, LU, and setting, LD, and the corresponding azimuths (radians), AU and AD, for a place on the surface of the Earth with geographical latitude GP (radians; N positive) where there is a vertical shift in the apparent position of the object at the horizon relative to the horizontal plane through the observer of DI (radians).

Set FL(4) to 0 on first call and whenever a new value of GP is specified. FL(4) is set to 1 by the routine.

Error status on return is indicated by ER(3) as follows:

ER(3) = 0 : normal return

ER(3) = 1 : object never rises

ER(3) = −1 : object is circumpolar.

```
                          3097  REM
                          3098  REM        Subroutine RISET
                          3099  REM
```

3100 Don't recalculate these values if they have
not changed
```
                          3100  IF FL(4)=1 THEN GOTO 3115
```

3105 Calculate trigonometric values just
once . . .
```
                          3105  CF=COS(GP) : SF=SIN(GP)
```

3110 . . . and set value of TP = 2π
```
                          3110  FL(4)=1 : TP=6.283185308
```

3115 Calculate these trigonometric values just
once also
```
                          3115  SY=SIN(Y) : CY=COS(Y) : SD=SIN(DI)
```

3120 CH is cos (hour angle) at horizon
```
                          3120  CD=COS(DI) : CH=-(SD+(SF*SY))/(CF*CY)
```

3125 CH more negative than −1 indicates star
never sets
```
                          3125  IF CH<-1.0 THEN GOTO 3190
```

3130 CH greater than +1 indicates that it never
rises
```
                          3130  IF CH>1.0 THEN GOTO 3195
```

3135 Rising and setting both occur, so . . .
```
                          3135  CA=(SY+(SD*SF))/(CD*CF)
                          3140  H=FNC(CH) : AU=FNC(CA) : ER(3)=0
                          3145  B=FND(H)/15.0 : A=FND(X)/15.0
                          3150  LU=24.0+A-B : LD=A+B : AD=TP-AU
```

3155 . . . find local sidereal times of rising
(LU) . . .
```
                          3155  C=LU : D=24.0 : GOSUB 3175 : LU=C
```

3160 . . . and setting (LD) in range 0–24 hours,
and . . .
```
                          3160  C=LD : GOSUB 3175 : LD=C
```

3165 . . . azimuths of rising (AU) and setting
(AD) . . .
```
                          3165  C=AU : D=TP : GOSUB 3175 : AU=C
```

3170 . . . in range 0–2π radians
```
                          3170  C=AD : GOSUB 3175 : AD=C : RETURN
```

3175 Subroutine to return value of C reduced
to . . .
```
                          3175  IF C<0.0 THEN C=C+D : GOTO 3175
```

3180 . . . the interval 0–D
```
                          3180  IF C>D THEN C=C-D : GOTO 3180
                          3185  RETURN
```

3190 Error returns after setting error flag
ER(3) . . .
```
                          3190  ER(3)=-1 : PRINT "** circumpolar **" : RETURN
```

3195 . . . and displaying a message
```
                          3195  ER(3)=1 : PRINT "** never rises **" : RETURN
```

Notes	Line	Code
	1	REM
	2	REM Handling program HRISET
	3	REM
	5	DIM FL(20),SW(20),ER(20)
10 FNI is needed by JULDAY	10	DEF FNI(W)=SGN(W)*INT(ABS(W))
15 FNM converts degrees to radians . . .	15	DEF FNM(W)=1.745329252E-2*W
20 . . . and FND converts radians to degrees	20	DEF FND(W)=5.729577951E1*W
25 FNS returns the inverse sine . . .	25	DEF FNS(W)=ATN(W/(SQR(1-W*W)+1E-20))
30 . . . and FNC the inverse cosine	30	DEF FNC(W)=1.570796327-FNS(W)
	35	PRINT : PRINT : FL(4)=0 : FL(3)=0
	40	PRINT "The circumstances of rising and setting"
	45	PRINT "---------------------------------------"
	50	PRINT : PRINT
55 FL(4)=1 indicates that we are in the same place as before	55	IF FL(4)=1 THEN GOTO 110
60 Get a new place . . .	60	Q$="Geog. longitude (D,M,S; W neg.) .... "
	65	PRINT Q$; : INPUT XD,XM,XS
70 . . . GL is geographical longitude in degrees (via MINSEC)	70	SW(1)=-1 : GOSUB 1000 : GL=X
	75	Q$="Geog. latitude (D,M,S; S neg.) ..... "
	80	PRINT Q$; : INPUT XD,XM,XS
85 . . . GP the latitude in radians	85	SW(1)=-1 : GOSUB 1000 : GP=FNM(X)
	90	Q$="Daylight saving (H ahead of zone t) "
95 . . . DS the number of hours added for daylight saving	95	PRINT Q$; : INPUT DS
105 . . . TZ is the time zone (hours ahead of UT)	100	Q$="Time zone (hours; West negative) ... "
	105	PRINT Q$; : INPUT TZ
110 FL(3)=1 indicates that we have the same date as before	110	IF FL(3)=1 THEN GOTO 125
120 Get a new date	115	Q$="Calendar date (D,M,Y) .... "
	120	PRINT : PRINT Q$; : INPUT DY,MN,YR
125 Now get the astronomical position of the object . . .	125	Q$="Right ascension (H,M,S) .... "
135 . . . right ascension in radians	130	PRINT : PRINT Q$; : INPUT XD,XM,XS
	135	SW(1)=-1 : GOSUB 1000 : XA=FNM(X)*15.0
	140	Q$="Declination (D,M,S) .... "
	145	PRINT Q$; : INPUT XD,XM,XS
150 . . . declination in radians	150	GOSUB 1000 : Y=FNM(X)
	155	Q$="Vertical displacement (D,M,S) ...... "
	160	PRINT Q$; : INPUT XD,XM,XS
165 . . . and its vertical displacement in radians	165	GOSUB 1000 : DI=FNM(X) : X=XA
170 Call RISET . . .	170	GOSUB 3100
175 . . . and check for errors: ER(3)=0 if none	175	IF ER(3)<>0 THEN GOTO 360
180 Display the results . . .	180	PRINT : PRINT "Circumstances of rising:"
	185	PRINT "------------------------" : PRINT
190 Convert local sidereal time of rising to H,M,S . . .	190	X=LU : NC=9 : SW(1)=1 : GOSUB 1000
200 . . . and display it neatly (columns aligned)	195	Q$="Local sidereal time (H,M,S) ...."
	200	PRINT Q$+" "+MID$(OP$,4,12)
205 Now find the corresponding Universal Time etc. by calling TIME	205	SW(2)=-1 : TM=LU : GOSUB 1300
210 ER(1)=1 indicates impossible date	210	IF ER(1)=1 THEN GOTO 360

215	Display all the other times . . .	215	`Q$="Greenwich sidereal time (H,M,S) .. "`
220	. . . neatly	220	`PRINT Q$+"    "+SG$`
		225	`Q$="Universal time (H,M,S) .......... "`
		230	`PRINT Q$+"    "+UT$`
		235	`Q$="...may be in error by up to 4 minutes"`
240	ER(2)=1 indicates that the conversion is ambiguous	240	`IF ER(2)=1 THEN PRINT Q$`
		245	`Q$="Local civil time (H,M,S) ........ "`
		250	`PRINT Q$+"    "+TL$`
255	Now display the azimuth in D,M,S format	255	`X=FND(AU) : SW(1)=1 : GOSUB 1000`
		260	`Q$="Azimuth (D,M,S; zero is North) ... "`
		265	`PRINT Q$+OP$`
270	Repeat the above for setting	270	`PRINT : PRINT "Circumstances of setting:"`
		275	`PRINT "------------------------" : PRINT`
		280	`X=LD : NC=9 : SW(1)=1 : GOSUB 1000`
		285	`Q$="Local sidereal time (H,M,S) ......."`
		290	`PRINT Q$+"    "+MID$(OP$,4,12)`
295	Call TIME for the other times (Universal Time etc.)	295	`SW(2)=-1 : TM=LD : GOSUB 1300`
300	ER(1)=1 indicates impossible date	300	`IF ER(1)=1 THEN GOTO 360`
305	Display all the times . . .	305	`Q$="Greenwich sidereal time (H,M,S) .. "`
310	. . . neatly	310	`PRINT Q$+"    "+SG$`
		315	`Q$="Universal time (H,M,S) .......... "`
		320	`PRINT Q$+"    "+UT$`
		325	`Q$="...may be in error by up to 4 minutes"`
330	ER(2)=1 indicates that the conversion was ambiguous	330	`IF ER(2)=1 THEN PRINT Q$`
		335	`Q$="Local civil time (H,M,S) ........ "`
		340	`PRINT Q$+"    "+TL$`
345	Display the azimuth of setting in D,M,S format	345	`X=FND(AD) : SW(1)=1 : GOSUB 1000`
		350	`Q$="Azimuth (D,M,S; zero is North) ... "`
		355	`PRINT Q$+OP$`
360	Another calculation?	360	`Q$="Again (Y or N) .................. "`
365	Call YESNO to see	365	`PRINT : GOSUB 960`
		370	`IF E=0 THEN STOP`
375	If FL(4)=0 then we have not yet made a calculation	375	`IF FL(4)=0 THEN PRINT : GOTO 55`
		380	`Q$="Same place (Y or N) ............. "`
385	Set FL(4)=0 for a new place	385	`GOSUB 960 : FL(4)=E`
		390	`Q$="Same date (Y/N) ................. "`
395	Set FL(3)=0 for a new date	395	`GOSUB 960 : FL(3)=E : PRINT : GOTO 55`

```
INCLUDE YESNO, MINSEC, JULDAY, TIME, RISET
```

3100 RISET

Example

```
The circumstances of rising and setting
----------------------------------------


Geog. longitude (D,M,S; W neg.) .... ? 62,10,12
Geog. latitude (D,M,S; S neg.) ..... ? -20,0,3.4
Daylight saving (H ahead of zone t) ? 0
Time zone (hours; West negative) ... ? 4

Calendar date (D,M,Y) .............. ? 23,9,1992

Right ascension (H,M,S) ............ ? 12,16,0
Declination (D,M,S) ................ ? 14,34,0
Vertical displacement (D,M,S) ...... ? 0,34,0

Circumstances of rising:
------------------------

Local sidereal time (H,M,S) .......    6 35 12.47
Greenwich sidereal time (H,M,S) ..    2 26 31.67
Universal time (H,M,S) ...........    2 17 47.81
Local civil time (H,M,S) .........    6 17 47.81
Azimuth (D,M,S; zero is North) ...  + 74 41 19.57

Circumstances of setting:
-------------------------

Local sidereal time (H,M,S) .......   17 56 47.53
Greenwich sidereal time (H,M,S) ..   13 48  6.73
Universal time (H,M,S) ...........   13 37 31.22
Local civil time (H,M,S) .........   17 37 31.22
Azimuth (D,M,S; zero is North) ...  +285 18 40.43

Again (Y or N) ................... ? Y
Same place (Y or N) .............. ? Y
Same date (Y/N) .................. ? Y


Right ascension (H,M,S) ............ ? 6,36,7
Declination (D,M,S) ................ ? 87,21,10
Vertical displacement (D,M,S) ...... ? 0,0,0
** never rises **

Again (Y or N) ................... ? Y
Same place (Y or N) .............. ? Y
Same date (Y/N) .................. ? Y


Right ascension (H,M,S) ............ ? 6,36,7
Declination (D,M,S) ................ ? -87,0,0
Vertical displacement (D,M,S) ...... ? 0,0,0
** circumpolar **

Again (Y or N) ................... ? N
```

3300 ANOMALY

This routine finds the values of the eccentric and true anomalies in elliptical motion, given the mean anomaly and the eccentricity.

The starting point for calculations involving a body in an elliptical orbit is often the *mean anomaly*, AM. This is the angle moved by a fictitious body in a circular orbit of the same period as the real body, the angle being reckoned in the same sense as the direction of motion of the real body from the point of closest approach (the *periapsis*). The quantity needed is the *true anomaly*, AT, which measures the angle moved by the real body since periapsis, and it is related to AM through the *eccentric anomaly*, AE, by Kepler's equation.

$$AE - (EC \times SIN(AE)) = AM,$$

where EC is the eccentricity of the orbit. Unfortunately, this equation is not easily solved, but the solution can be approximated by a trigonometric expansion called the *equation of the centre*. If EC is less than about 0.1 and high precision is not required, the first term of the expansion may suffice, giving

$$AT = AM + (2 \times EC \times SIN(AM)),$$

where AT and AM are expressed in radians.

For more accurate work, the equation must be solved explicitly for AE, and then AT calculated from

$$TAN(AT/2) = ((1 + EC)/(1 - EC))^{0.5} \times TAN(AE/2).$$

The routine given here solves Kepler's equation by an iterative method in which an approximate solution for AE is repeatedly refined until the error between $(AE - (EC \times SIN(AE)))$ and AM is less than a given error (10^{-6} radians). The refinement, D, made to AE in each step is given by

$$D = (AE - (EC \times SIN(AE)) - AM)/(1 - (EC \times COS(AE))),$$

so that the new value of AE becomes AE − D. The routine will always converge rapidly for EC < 1 from any starting value, though the more accurate it is, and the smaller the value of EC, the fewer the number of iterations needed. Here, the starting point is always taken as AE = AM, the exact solution when EC = 0 (circular orbit).

The allowed error in the iteration is set in line 3310 to 10^{-6} radians. If your machine uses more than four bytes to represent real numbers, you can make the error smaller to increase the precision with which AT is calculated from AM.

3300 ANOMALY

However, if this value is made too small, rounding errors in the BASIC arithmetic routines may be sufficient to ensure that the IF condition of line 3310 is never satisfied and then execution will get stuck in an infinite loop. If a program using ANOMALY appears to go to sleep, try increasing the allowed error in line 3310.

Details of ANOMALY
Called by GOSUB 3300.
Calculates AT, the true anomaly, and AE, the eccentric anomaly in elliptical motion given AM, the mean anomaly, and EC (<1), the eccentricity. All angles in *radians*.

```
3297   REM
3298   REM        Subroutine ANOMALY
3299   REM
```

3300	TP is 2π; initial settings for iterative loop	`3300   TP=6.283185308 : M=AM-TP*INT(AM/TP) : AE=M`
3305	Find error between current estimate of M and its true value	`3305   D=AE-(EC*SIN(AE))-M`
3310	Loop until the error is less than 1E–6 (radians)	`3310   IF ABS(D)<1E-6 THEN GOTO 3320`
3315	Refine AE, and loop again	`3315   D=D/(1.0-(EC*COS(AE))) : AE=AE-D : GOTO 3305`
3320	Calculate the true anomaly, AT	`3320   A=SQR((1.0+EC)/(1.0-EC))*TAN(AE/2.0)`
3325	Return from ANOMALY	`3325   AT=2.0*ATN(A) : RETURN`

```
1      REM
2      REM        Handling program HANOMALY
3      REM
```

		`5      DIM SW(20)`
10	Degrees to radians conversion . . .	`10     DEF FNM(W)=1.745329252E-2*W`
15	. . . and radians to degrees	`15     DEF FND(W)=5.729577951E1*W`
		`20     PRINT : PRINT`
		`25     PRINT "Solving Kepler's equation"`
		`30     PRINT "------------------------"`
		`35     PRINT`
40	Get the mean anomaly (AM) . . .	`40     Q$="Mean anomaly (D,M,S) ......... "`
		`45     PRINT Q$; : INPUT XD,XM,XS`
50	. . . in radians via MINSEC and FNM	`50     SW(1)=-1 : GOSUB 1000 : AM=FNM(X)`
55	Get the eccentricity (EC) . . .	`55     Q$="Eccentricity (0 to 1) ......... "`
		`60     PRINT Q$; : INPUT EC`
65	. . . ensuring that it is within the allowed bounds	`65     IF EC<0 OR EC>1.0 THEN GOTO 60`
70	Call ANOMALY . . .	`70     GOSUB 3300 : PRINT : X=FND(AE)`
75	. . . and display results in D,M,S forms	`75     NC=9 : SW(1)=1 : GOSUB 1000`
		`80     Q$="Eccentric anomaly (D,M,S) ..... "`
		`85     PRINT Q$+OP$ : X=FND(AT) : GOSUB 1000`
		`90     Q$="True anomaly (D,M,S) .......... "`
		`95     PRINT Q$+OP$`
100	Another go?	`100    Q$="Again (Y or N) ............... "`
105	Call YESNO to get a response	`105    PRINT : GOSUB 960`
		`110    IF E=1 THEN GOTO 35`
		`115    STOP`

```
INCLUDE YESNO, MINSEC, ANOMALY
```

3300 ANOMALY

Example

```
Solving Kepler's equation
-------------------------

Mean anomaly (D,M,S) .......... ? 220,23,10
Eccentricity (0 to 1) ......... ? 0.0167183

Eccentric anomaly (D,M,S) .....   +219 46 23.76
True anomaly (D,M,S) ..........   -140 50  8.41

Again (Y or N) ................ ? Y

Mean anomaly (D,M,S) .......... ? 45,0,0
Eccentricity (0 to 1) ......... ? 0.96629

Eccentric anomaly (D,M,S) .....   + 99 35 25.95
True anomaly (D,M,S) ..........   +167 22 11.55

Again (Y or N) ................ ? N
```

Higher accuracy can be obtained (if allowed by the precision of the number-representation of your machine) by decreasing the error threshold in line 3310 (at present set to 1E–6).

3400 SUN

This routine calculates the true ecliptic longitude and geocentric distance of the Sun at any given instant.

The Sun's celestial coordinates are determined by its position in its apparent orbit of the Earth. Although the Earth actually orbits the Sun, it seems to us that it is the Sun which is in orbit about the Earth, and for the purposes of calculating the Sun's position it is convenient to regard this as the case. We can then compute the ecliptic longitude of the Sun for a given instant and, since the Sun–Earth orbit defines the plane of the ecliptic, we can take the ecliptic latitude to be 0. The right ascension and declination can be found by using routine EQECL (2000) followed by EQHOR (1500) if we wish to know the Sun's azimuth and altitude as well.

Subroutine SUN calculates the true geocentric ecliptic longitude of the Sun, SR, and its distance from the Earth, RR, at the instant given by UT (Universal Time, hours), DY (integer days), MN (months), and YR (years). The time may also be expressed as a fraction of a day and added to the value of DY, in which case UT must be set to 0. The true ecliptic latitude may be assumed to be 0 (always correct to within 1.2 arcseconds). SR is the longitude referred to the mean equinox of date and is the quantity required in the calculation of geocentric planetary positions. However, if the *apparent* longitude of the Sun is needed, it is necessary to correct for nutation and aberration. Routine NUTAT (1800) calculates DP, the correction to be added to SR for nutation (and also DO to be added to the obliquity of the ecliptic), while aberration amounts to a fixed correction of 5.69×10^{-3} degrees. Thus

   SRA = SR + DP − 5.69E−3,

where SRA is the Sun's apparent longitude and all the quantities are expressed in degrees. The correction for aberration is due to the speed of the Earth in its orbit. Since light does not travel at infinite speed, the motion of the Earth causes the apparent direction of a celestial body to be shifted slightly from its true direction, just as rain falling vertically downwards appears to come at an angle to a cyclist moving through it. The effect is small, amounting to a maximum shift of 20.4 arcseconds (5.69×10^{-3} degrees). The correction for the Sun is a constant because the Earth is always travelling more or less at right angles to the Sun–Earth line (a nearly circular orbit). The correction for a star, however, would depend on its position in relation to the Sun.

SUN makes calls to ANOMALY (3300) to solve Kepler's equation, and to JULDAY (1100) to convert the date to Julian days. It makes no change to FL(1) before doing so. Hence, if FL(1) = 1, the current value of DJ is assumed to be correct.

Formulae

$SR = AT + L - M1 + D2$

$RR = (1.0000002 \times (1 - (EC \times COS(EA)))) + D3$

$L = 2.7969668E2 + (3.600076892E4 \times T) + (3.025E{-}4 \times T^2)$

$M1 = 3.5847583E2 + (3.599904975E4 \times T) - (1.54E{-}4 \times T^2)$
$\qquad - (3.3E{-}6 \times T^3)$

$EC = 1.675104E{-}2 - (4.18E{-}5 \times T) - (1.26E{-}7 \times T^2)$

AT = true anomaly

EA = eccentric anomaly

$M1$ = mean anomaly

L = mean longitude

$D3$ = correction to RR for perturbations

$D2$ = correction to SR for perturbations

$T = DJ/36525$

DJ = Julian days since 1900 January 0.5

Most of the expressions involving powers of T result in very large numbers which have to be reduced near to the range 0–360 degrees. Better accuracy is obtained if this reduction is carried out before multiplying by T. Hence the preferred form for L is

$L = 2.7969668E2 + (3.025E{-}4 \times T^2) + B$

$B = 360 \times (A - INT(A))$

$A = 1.000021359E2 \times T$

INT can be either the least-integer or the truncated-integer function. Similar forms are adopted for the other expressions used by SUN (see the program listing for details).

Details of SUN

Called by GOSUB 3400.

Calculates SR, the true geocentric longitude of the Sun for the mean equinox of date (*radians*), and RR, the Sun–Earth distance (*AU*), for the instant defined by UT (Universal Time, hours), DY (integer days), MN (months), and YR (years). The Universal Time may also be included by adding it as a fraction of a day to DY, and setting UT to 0.

Other routines called: JULDAY (1100), ANOMALY (3300).

```
3397 REM
3398 REM        Subroutine SUN
3399 REM
```

3400	Call JULDAY for DJ; T is instant expressed in centuries	`3400 GOSUB 1100 : T=(DJ/36525.0)+(UT/8.766E5)  : T2=T*T`
3405	Calculate . . .	`3405 A=1.000021359E2*T : B=360.0*(A-INT(A))`
3410	. . . Sun's mean longitude, L (degrees)	`3410 L=2.7969668E2+3.025E-4*T2+B`
		`3415 A=9.999736042E1*T : B=360.0*(A-INT(A))`
3420	. . . Sun's mean anomaly, M1 (degrees)	`3420 M1=3.5847583E2-(1.5E-4+3.3E-6*T)*T2+B`
3425	. . . Earth's orbital eccentricity	`3425 EC=1.675104E-2-4.18E-5*T-1.26E-7*T2`
3430	Call ANOMALY to find true anomaly, AT (radians)	`3430 AM=FNM(M1) : GOSUB 3300`
3435	Find the various arguments . . .	`3435 A=6.255209472E1*T : B=360.0*(A-INT(A))`
		`3440 A1=FNM(153.23+B)`
		`3445 A=1.251041894E2*T : B=360.0*(A-INT(A))`
		`3450 B1=FNM(216.57+B)`
		`3455 A=9.156766028E1*T : B=360.0*(A-INT(A))`
		`3460 C1=FNM(312.69+B)`
		`3465 A=1.236853095E3*T : B=360.0*(A-INT(A))`
		`3470 D1=FNM(350.74-1.44E-3*T2+B)`
		`3475 E1=FNM(231.19+20.2*T)`
		`3480 A=1.831353208E2*T : B=360.0*(A-INT(A))`
		`3485 H1=FNM(353.4+B)`
3490	. . . so that the corrections to . . .	`3490 D2=1.34E-3*COS(A1)+1.54E-3*COS(B1)+2E-3*COS(C1)`
3495	. . . the Sun's geocentric longitude, D2 . . .	`3495 D2=D2+1.79E-3*SIN(D1)+1.78E-3*SIN(E1)`
3500	. . . and radius vector, D3, can be found	`3500 D3=5.43E-6*SIN(A1)+1.575E-5*SIN(B1)`
		`3505 D3=D3+1.627E-5*SIN(C1)+3.076E-5*COS(D1)`
		`3510 D3=D3+9.27E-6*SIN(H1)`
3515	Now find Sun's geocentric longitude, SR (radians) . . .	`3515 SR=AT+FNM(L-M1+D2) : TP=6.283185308`
3520	. . . and Sun–Earth distance, RR (AU)	`3520 RR=1.0000002*(1.0-EC*COS(AE))+D3`
3525	Make sure that SR is in range 0–2π (radians)	`3525 IF SR<0 THEN SR=SR+TP : GOTO 3525`
		`3530 IF SR>TP THEN SR=SR-TP : GOTO 3530`
3535	Return from SUN	`3535 RETURN`

	1	REM
	2	REM Handling program HSUN
	3	REM
	5	DIM FL(20),ER(20),SW(20)
10	FNI is needed by JULDAY ...	10	DEF FNI(W)=SGN(W)*INT(ABS(W))
15	Converts degrees to radians ...	15	DEF FNM(W)=1.745329252E-2*W
20	... and radians to degrees	20	DEF FND(W)=5.729577951E1*W
25	FNS returns inverse sine	25	DEF FNS(W)=ATN(W/(SQR(1-W*W)+1E-20))
	30	PRINT : PRINT : SP=0 : SH=0
	35	PRINT "The position of the Sun"
	40	PRINT "----------------------" : PRINT : PRINT
45	SP is a local switch meaning 'same place' if set to 1	45	IF SP=1 THEN GOTO 70
50	Get location details	50	Q$="Daylight saving (H ahead of zone t) "
	55	PRINT Q$; : INPUT DS
	60	Q$="Time zone (hours; West negative) ... "
	65	PRINT Q$; : INPUT TZ
70	FL(1) is set to 1 if the date has not changed	70	IF FL(1)=1 THEN GOTO 85
75	Get a new date	75	Q$="Calendar date (D,M,Y) .... "
	80	PRINT : PRINT Q$; : INPUT DY,MN,YR
85	SH is a local switch meaning 'same time' if set to 1	85	IF SH=1 THEN GOTO 105
90	Get a new time ...	90	Q$="Local civil time (H,M,S) .... "
	95	PRINT : PRINT Q$; : INPUT XD,XM,XS
100	... in decimal hours (via MINSEC)	100	SW(1)=-1 : GOSUB 1000 : TM=X
105	Calculate UT, and call SUN	105	UT=TM-DS-TZ : GOSUB 3400
110	Was the date impossible?	110	IF ER(1)=1 THEN GOTO 225
115	Display the results ...	115	PRINT : PRINT "Geocentric coordinates"
	120	PRINT "----------------------" : PRINT
125	... in format D,M,S, or H,M,S ...	125	SW(1)=1 : NC=9 : X=FND(SR) : GOSUB 1000
130	... true values ...	130	Q$="True ecliptic longitude (D,M,S) .... "
	135	PRINT Q$+OP$
	140	Q$="True distance (AU) .... "
145	correct to five decimal places ...	145	PRINT Q$+" "+STR$(INT(1E5*RR+0.5)/1E5)
150	... call EQECL to convert to right ascension and declination	150	X=SR : Y=0 : SW(3)=-1 : GOSUB 2000
	155	P=FND(P) : Q=FND(Q) : X=P/15.0 : GOSUB 1000
	160	Q$="True right ascension (H,M,S) .... "
	165	PRINT Q$+" "+MID$(OP$,4,12) : X=Q : GOSUB 1000
	170	Q$="True declination (D,M,S) .... "
	175	PRINT Q$+OP$: PRINT : GOSUB 1800
180	Correct for nutation and aberration ...	180	AL=FND(SR)+DP-5.69E-3 : X=AL : GOSUB 1000
185	... to get apparent values ...	185	Q$="Apparent ecliptic longitude (D,M,S) "
	190	PRINT Q$+OP$
195	... call EQECL to convert to right ascension and declination ...	195	X=FNM(AL) : Y=0 : SW(3)=-1 : GOSUB 2000
200	... display in H,M,S format (via MINSET) ...	200	P=FND(P) : Q=FND(Q) : X=P/15.0 : GOSUB 1000
210	... neatly	205	Q$="Apparent right ascension (H,M,S) ... "
	210	PRINT Q$+" "+MID$(OP$,4,12) : X=Q : GOSUB 1000
	215	Q$="Apparent declination (D,M,S) .... "
	220	PRINT Q$+OP$
225	Another go?	225	Q$="Again (Y or N) .... "
230	Ask YESNO to see	230	PRINT : GOSUB 960
	235	IF E=0 THEN STOP
	240	Q$="Same place (Y or N) .... "

118

245	SP=0 for a new time zone or daylight saving correction	245	GOSUB 960 : SP=E
		250	Q$="Same date (Y or N) .... "
255	Flags set to 0 for new date: 1: JULDAY, 6: NUTAT, 7: EQECL	255	GOSUB 960 : FL(1)=E : FL(7)=E : FL(6)=E
		260	Q$="Same time (Y or N) .... "
265	SH=0 for new time	265	GOSUB 960 : SH=E : PRINT : GOTO 45

```
INCLUDE YESNO, MINSEC, JULDAY, OBLIQ, NUTAT, EQECL,
        ANOMALY, SUN
```

Example

```
The position of the Sun
-----------------------


Daylight saving (H ahead of zone t)  ? 0
Time zone (hours; West negative) ... ? -5

Calendar date (D,M,Y) .............. ? 23,8,1984

Local civil time (H,M,S) ........... ? 19,0,0

Geocentric coordinates
----------------------

True ecliptic longitude (D,M,S) ....   +151  0 47.14
True distance (AU) .................    1.01099
True right ascension (H,M,S) .......      10 12 13.58
True declination (D,M,S) ...........   + 11  6 56.69

Apparent ecliptic longitude (D,M,S)    +151  0 12.39
Apparent right ascension (H,M,S) ...      10 12 11.37
Apparent declination (D,M,S) .......   + 11  7  9.01

Again (Y or N) ..................... ? Y
Same place (Y or N) ................ ? N
Same date (Y or N) ................. ? N
Same time (Y or N) ................. ? N

Daylight saving (H ahead of zone t)  ? 0
Time zone (hours; West negative) ... ? 0

Calendar date (D,M,Y) .............. ? 24,8,1984

Local civil time (H,M,S) ........... ? 0,0,0

Geocentric coordinates
----------------------

True ecliptic longitude (D,M,S) ....   +151  0 47.14
True distance (AU) .................    1.01099
True right ascension (H,M,S) .......      10 12 13.58
True declination (D,M,S) ...........   + 11  6 56.71

Apparent ecliptic longitude (D,M,S)    +151  0 12.48
Apparent right ascension (H,M,S) ...      10 12 11.38
Apparent declination (D,M,S) .......   + 11  7  9.00

Again (Y or N) ..................... ? N
```

3400 SUN

The Astronomical Almanac gives the following values:

Date	Apparent right ascension	Apparent declination	True distance
24th August 1984	10 12 11.46	11 07 09.1	1.0109976

This handling program displays an ephemeris of the Sun, that is a list of its positions calculated at regular intervals.

Notes	HSUNEP

	1	REM
	2	REM Handling program HSUNEP
	3	REM
10 FNI is needed by JULDAY...	5	DIM FL(20),ER(20),SW(20)
15 ...and FNL by CALDAY	10	DEF FNI(W)=SGN(W)*INT(ABS(W))
20 This function converts degrees to radians...	15	DEF FNL(W)=FNI(W)+FNI((SGN(W)-1.0)/2.0)
25 ...and this one converts radians to degrees	20	DEF FNM(W)=1.745329252E-2*W
30 This returns the inverse sine function	25	DEF FND(W)=5.729577951E1*W
	30	DEF FNS(W)=ATN(W/(SQR(1-W*W)+1E-20))
	35	PRINT : PRINT : PRINT "Solar ephemeris"
	40	PRINT "--------------" : PRINT
	45	PRINT
50 Get the starting date for the ephemeris	50	Q$="Starting date (D,M,Y) .... "
	55	PRINT : PRINT Q$; : INPUT DA,MA,YA
60 How often do we want a new position to be calculated?	60	Q$="Step interval (days) .... "
	65	PRINT Q$; : INPUT SD
70 ...and how long should the list be?	70	Q$="... for how many steps .... "
	75	PRINT Q$; : INPUT NS
80 Do we want to display apparent coordinates...	80	Q$="Apparent coordinates (Y or N) ...... "
85 ...YESNO will get the answer...	85	PRINT : GOSUB 960
90 (IS is the local switch for this question)	90	IF E=1 THEN IS=0 : PRINT : GOTO 110
95 ...or true coordinates?	95	Q$="True coordinates (Y or N) .... "
	100	GOSUB 960 : IS=1 : PRINT
105 If neither, skip to the end	105	IF E=0 THEN GOTO 245
110 Display the appropriate heading...	110	Q$="Apparent geocentric coordinates at 0h UT"
	115	IF IS=0 THEN PRINT Q$: GOTO 125
	120	PRINT "True geocentric coordinates at 0h UT"
	125	PRINT
130 ...and label all the columns of the list	130	PRINT TAB(5);"DATE";TAB(19);"LONGITUDE";
	135	PRINT TAB(33);"RIGHT ASCENSION";TAB(51);"DECLINATION"
	140	PRINT TAB(22);"DMS";TAB(38);"HMS";TAB(55);"DMS"
	145	PRINT "---";
	150	PRINT "-------------------"
	155	PRINT

☞

120

160	Get the Julian Date, D0, of the start . . .	160	`DY=DA : MN=MA : YR=YA : FL(1)=0`
165	. . . by calling JULDAY	165	`GOSUB 1100 : DO=DJ`
170	Come on, let's not begin on an impossible date!	170	`IF ER(1)=1 THEN GOTO 245`
175	Now loop to make the list (index II) . . .	175	`FOR II=0 TO NS-1 : FL(2)=0`
180	. . . get the next value of DJ; find the calendar date . . .	180	`DJ=D0+II*SD : GOSUB 1200 : UT=0 : GOSUB 3400`
185	. . . from CALDAY; call SUN; call NUTAT for nutation . . .	185	`FL(6)=0 : GOSUB 1800`
190	. . . convert longitude to degrees . . .	190	`SW(1)=1 : NC=9 : AL=FND(SR)`
195	. . . and correct for nutation and aberration if apparent . . .	195	`IF IS=0 THEN AL=AL+DP-5.69E-3`
200	. . . call MINSEC for longitude in D,M,S form . . .	200	`X=AL : GOSUB 1000 : L$=OP$ : X=FNM(AL)`
205	. . . call EQECL to find right ascension and declination	205	`Y=0 : SW(3)=-1 : FL(5)=0 : FL(7)=0 : GOSUB 2000`
210	. . . in degrees (hours) . . .	210	`P=FND(P) : Q=FND(Q) : X=P/15.0`
215	. . . call MINSEC for D/H,M,S format . . .	215	`GOSUB 1000 : R$=MID$(OP$,4,12)`
		220	`X=Q : GOSUB 1000 : D$=OP$`
225	. . . and display the results neatly . . .	225	`PRINT DT$;TAB(15);L$;TAB(33);R$;TAB(48);D$`
230	. . . end of loop; repeat for NS steps	230	`NEXT II`
235	Finish off the list neatly	235	`PRINT "---------------------------------------------";`
		240	`PRINT "---------------------"`
245	Some more?	245	`Q$="Again (Y or N) ..................... "`
250	YESNO gets the answer	250	`PRINT : GOSUB 960`
		255	`IF E=0 THEN STOP`
260	Repeat this, or get fresh dates etc.?	260	`Q$="New dates (Y or N) ................. "`
		265	`GOSUB 960`
		270	`IF E=1 THEN GOTO 50`
		275	`GOTO 80`

```
INCLUDE YESNO, MINSEC, JULDAY, CALDAY, OBLIQ, NUTAT, EQECL,
        ANOMALY, SUN
```

3400 SUN

Example

```
Solar ephemeris
---------------



Starting date (D,M,Y) ............. ? 23,8,1984
Step interval (days) .............. ? 1
... for how many steps ............ ? 4

Apparent coordinates (Y or N) ...... ? Y

Apparent geocentric coordinates at 0h UT
```

DATE	LONGITUDE DMS	RIGHT ASCENSION HMS	DECLINATION DMS
23 8 1984	+150 2 20.62	10 8 30.56	+ 11 27 35.09
24 8 1984	+151 0 12.48	10 12 11.38	+ 11 7 9.00
25 8 1984	+151 58 5.76	10 15 51.77	+ 10 46 32.38
26 8 1984	+152 56 0.59	10 19 31.75	+ 10 25 45.48

```
Again (Y or N) .................... ? Y
New dates (Y or N) ................ ? N

Apparent coordinates (Y or N) ...... ? N
True coordinates (Y or N) ......... ? Y

True geocentric coordinates at 0h UT
```

DATE	LONGITUDE DMS	RIGHT ASCENSION HMS	DECLINATION DMS
23 8 1984	+150 2 55.38	10 8 32.77	+ 11 27 22.86
24 8 1984	+151 0 47.14	10 12 13.58	+ 11 6 56.71
25 8 1984	+151 58 40.36	10 15 53.96	+ 10 46 20.01
26 8 1984	+152 56 35.19	10 19 33.94	+ 10 25 33.02

```
Again (Y or N) .................... ? N
```

3600 SUNRS

This routine calculates the universal and local civil times and azimuths of sunrise and sunset, or the beginning and end of civil, nautical, or astronomical twilight, at a given place anywhere on the Earth's surface for a given calendar date.

Routine RISET (3100) calculates the local sidereal times and azimuths of a celestial body given its right ascension and declination as input parameters. We may combine this routine with SUN (3400) and EQECL (2000) to calculate the right ascension and declination of the Sun on any given date, and hence the circumstances of sunrise and sunset. The procedure is complicated, however, by the fact that the Sun is not stationary in the sky with respect to the background of stars but moves by about 1 degree of ecliptic longitude per day. The calculated circumstances of rising and setting would be correct for an object at the Sun's position at the given instant, but would not correspond exactly with the observed circumstances of sunrise or sunset unless we had chosen the correct time already. We therefore have to adopt an iterative procedure, refining an initial crude estimate to achieve the required accuracy.

Routine SUNRS first calculates the Sun's right ascension and declination at local midday of the date in question, and hence the local sidereal times of rising and setting corresponding to the Sun's position at midday. The times are then corrected for the observer's longitude to find the Greenwich sidereal times, and converted to universal times using routine TIME. These times are already fair approximations to the actual times of rising and setting. The Sun's position is now recalculated for each of the times and is used to find better values for the circumstances of sunrise and sunset. The new values are accurate to about a second of time, much better than the uncertainties in atmospheric refraction and irregularities of the Earth's surface which conspire to alter the circumstances actually observed at any place by much more than this.

SUNRS calculates the universal and local civil times of rising and setting, returning the results in string form in the variables UU$ (UT), TU$ (local) for rising, and UD$ (UT), TD$ (local) for setting. The corresponding azimuths AU, AD, are also calculated. The routine can instead be used to find the times of the beginning and end of twilight, the beginning being returned in UU$, TU$ and the end in UD$, TD$. Switch SW(7) determines what sort of calcu-

3600 SUNRS

lation is performed. Set SW(7)=0 for sunrise and sunset; SW(7)=1 for civil twilight; SW(7)=2 for nautical twilight; SW(7)=3 for astronomical twilight.

The date is input as usual via the variables DY, MN, YR. The place is specified by its geographical longitude, GL, geographical latitude, GP, time zone, TZ, and daylight saving correction, DS. The routine calculates the circumstances of rising and setting appropriate to the Sun's upper limb, and adopts the standard value of horizontal atmospheric refraction of 34 arcminutes. It also assumes that the Sun's angular diameter is constant at 32 arcminutes 03 arcseconds and that the equatorial horizontal parallax is constant at 8.8 arcseconds.

Twilight is reckoned as the period of semi-darkness after sunset or before sunrise during which the Sun's zenith distance is between 90 degrees and some agreed amount. This is 96 degrees for civil twilight, 102 degrees for nautical twilight, and 108 degrees for astronomical twilight. SUNRS makes the calculations for twilight using the parameter DI of RISET, setting the appropriate values in lines 3605–3620.

Error conditions detected in SUNRS are reported by the error flags ER(4) and ER(5). Both flags are set to 0 on a normal return from the routine, but if the conversion between universal and sidereal times resulted in an ambiguity (see TIME), one or both of the flags is set to 1. Thus ER(4) = 1 if there is an ambiguity in the time of sunrise, and ER(5) = 1 if sunset. Note that SUNRS calls several other routines, including RISET. Its error flag, ER(3), will be set to 1 if the Sun never rises, and to −1 if it never sets (polar regions). If twilight lasts all night, the routine reacts as if the Sun never sets, and RISET will report the error message '** circumpolar **', while if there is no twilight, you'll get the message '** never rises **'. Remember to reset all the flags used by the relevant routines before you make repeated calls for different days or locations (see HSUNRS, lines 345 to the end).

Details of SUNRS

Called by GOSUB 3600.

Calculates the UT of sunrise, UU$, the local civil time of sunrise, TU$, the UT of sunset, UD$, and the local civil time of sunset, TD$, returning the times in string formats in those variables. The corresponding azimuths, AU and AD are also returned (*radians*). Alternatively, the routine calculates the beginning and end of twilight, returning the times in the same string variables as for rising (beginning of twilight) and setting (end of twilight). Input parameters are the date, DY (days), MN (months), YR (years), the geographical longitude, GL (*degrees*) and latitude (*radians*), the time zone, TZ (hours), the daylight saving correction, DS (hours), and the calculation type SW(7) as follows:

124

Set SW(7)=0 for sunrise and sunset.
Set SW(7)=1 for civil twilight.
Set SW(7)=2 for nautical twilight.
Set SW(7)=3 for astronomical twilight.
ER(4) is set to 1 by the routine if there is a possible error in the time of sunrise
or the start of twilight.
ER(5) is set to 1 by the routine if there is a possible error in the time of sunset
or the end of twilight.
Both ER(4) and ER(5) are set to 0 otherwise.
Other routines called: TIME (1300), NUTAT (1800), EQECL (2000), RISET
(3100), and SUN (3400).

```
3597 REM
3598 REM        Subroutine SUNRS
3599 REM
```

3600	Set values of displacement at horizon . . .	`3600 DI=1.454441E-2 : SW(2)=-1`
3605	. . . for sunrise and sunset . . .	`3605 IF SW(7)=0 THEN DI=1.454441E-2`
3610	. . . for civil twilight . . .	`3610 IF SW(7)=1 THEN DI=1.047198E-1`
3615	. . . for nautical twilight . . .	`3615 IF SW(7)=2 THEN DI=2.094395E-1`
3620	. . . for astronomical twilight	`3620 IF SW(7)=3 THEN DI=3.141593E-1`
3625	Find UT of local midday, call local routine . . .	`3625 UT=12.0+TZ+DS : GOSUB 3710`
3630	. . . to find local sidereal times of rising and setting	`3630 LA=LU : LB=LD`
3635	Return to base if any sort of error	`3635 IF ER(1)=1 OR ER(3)<>0 THEN RETURN`
3640	Find UT of rising using TIME . . .	`3640 TM=LA : GOSUB 1300 : GU=UT`
3645	. . . and of setting . . .	`3645 TM=LB : GOSUB 1300 : GD=UT`
3650	. . . and use these to find better times of rising . . .	`3650 UT=GU : GOSUB 3690 : LA=LU : AA=AU`

3660	. . . and setting using local routine	`3655 IF ER(3)<>0 THEN RETURN`
		`3660 UT=GD : GOSUB 3690 : LB=LD : AB=AD`
		`3665 IF ER(3)<>0 THEN RETURN`

3670	Now convert local sidereal times of sunrise . . .	`3670 TM=LA : GOSUB 1300 : ER(4)=ER(2)`
3680	. . . and sunset to UT and local civil time . . .	`3675 UU$=UT$ : TU$=TL$ : AU=AA` `3680 TM=LB : GOSUB 1300 : ER(5)=ER(2)`
3685	. . . output string formats, and return from SUNRS	`3685 UD$=UT$ : TD$=TL$ : AD=AB : RETURN`
3690	Save current DJ, and find whether local date . . .	`3690 DN=DJ : A=UT+TZ+DS`
3695	. . . is different from UT date. If it is, correct . . .	`3695 IF A>24 THEN DJ=DJ-1`
3700	. . . UT date before . . .	`3700 IF A<0 THEN DJ=DJ+1`
3705	. . . calling local routine. Restore DJ	`3705 GOSUB 3710 : DJ=DN : RETURN`
3710	Call SUN for solar longitude; call NUTAT for nutation	`3710 GOSUB 3400 : GOSUB 1800`
3715	Bad date?	`3715 IF ER(1)=1 THEN RETURN`
3720	Correct longitude for nutation and aberration	`3720 X=SR+FNM(DP-5.69E-3)`
3725	Call EQECL for sun right ascension and declination, which are then . . .	`3725 Y=0 : SW(3)=-1 : GOSUB 2000`
3730	. . . passed to RISET to get LST of rising and setting.	`3730 X=P : Y=Q : GOSUB 3100` `3735 RETURN`

```
1     REM
2     REM       Handling program HSUNRS
3     REM

5     DIM FL(20),ER(20),SW(20)
```

10   FNI needed by JULDAY

```
10    DEF FNI(W)=SGN(W)*INT(ABS(W))
```

15   FNM converts degrees to radians . . .

```
15    DEF FNM(W)=1.745329252E-2*W
```

20   . . . and FND converts radians to degrees

```
20    DEF FND(W)=5.729577951E1*W
```

25   FNS returns the inverse sine . . .

```
25    DEF FNS(W)=ATN(W/(SQR(1-W*W)+1E-20))
```

30   . . . and FNC returns the inverse cosine

```
30    DEF FNC(W)=1.570796327-FNS(W)

35    PRINT : PRINT
40    PRINT "Sunrise and sunset and twilight"
45    PRINT "-------------------------------" : PRINT
```

50   FL(4)=0 would indicate a new location

```
50    IF FL(4)=1 THEN GOTO 105
```

55   Get the location details . . .

```
55    Q$="Geog. longitude (D,M,S; W neg.) .... "
60    PRINT Q$; : INPUT XD,XM,XS
```

65   . . . longitude, GL, in degrees

```
65    SW(1)=-1 : GOSUB 1000 : GL=X
70    Q$="Geog. latitude (D,M,S; S neg.) ..... "
75    PRINT Q$; : INPUT XD,XM,XS
```

80   . . . latitude, GP, in radians

```
80    SW(1)=-1 : GOSUB 1000 : GP=FNM(X)
85    Q$="Daylight saving (H ahead of zone t) "
```

90   . . . daylight saving in hours (added to zone time)

```
90    PRINT Q$; : INPUT DS

95    Q$="Time zone (hours; West negative) ... "
```

100   . . . time zone in hours (added to UT)

```
100   PRINT Q$; : INPUT TZ
```

105   FL(3)=0 would indicate a new date

```
105   IF FL(3)=1 THEN GOTO 120
```

110   Get a new date

```
110   Q$="Calendar date (D,M,Y) ............. "
115   PRINT : PRINT Q$; : INPUT DY,MN,YR
```

120   What sort of calculation?

```
120   PRINT : PRINT "Calculation type:"
125   Q$="Sunrise and sunset (Y or N) ........ "
```

130   Call YESNO for yes or no answers

```
130   GOSUB 960
135   IF E=1 THEN SW(7)=0 : GOTO 190
140   Q$="Civil twilight (Y or N) ........... "
145   GOSUB 960
150   IF E=1 THEN SW(7)=1 : GOTO 190
155   Q$="Nautical twilight (Y or N) ........ "
160   GOSUB 960
165   IF E=1 THEN SW(7)=2 : GOTO 190
170   Q$="Astronomical twilight (Y or N) ..... "
175   GOSUB 960
180   IF E=1 THEN SW(7)=3 : GOTO 190
```

185   We get here if we have made no selection; skip to end

```
185   GOTO 345
```

190   Call SUNRS . . .

```
190   GOSUB 3600
```

195   . . . and skip to end if there were errors

```
195   IF ER(1)=1 OR ER(3)<>0 THEN GOTO 345
```

200   What sort of headline?

```
200   IF SW(7)>0 THEN GOTO 215
```

205   Display the appropriate headline

```
205   PRINT : PRINT "Circumstances of sunrise:"
210   PRINT "-------------------------" : GOTO 225
215   PRINT : PRINT "Morning twilight:"
220   PRINT "----------------"
```

225   Now display the results . . .

```
225   Q$="Universal time (H,M,S) ............ "
```

230   . . . they are already in string format

```
230   PRINT : PRINT Q$+UU$
235   Q$="...may be in error by up to 4 minutes"
```

240   Announce ambiguity (see TIME)

```
240   IF ER(4)=1 THEN PRINT Q$
```

☛

127

3600 SUNRS

	245 `Q$="Local civil time (H,M,S) .......... "`
	250 `PRINT Q$+TU$`
255 No azimuths if twilight	255 `IF SW(7)>0 THEN GOTO 285`
260 Convert to D,M,S format with FND and MINSEC...	260 `X=FND(AU) : SW(1)=1 : GOSUB 1000`
	265 `Q$="Azimuth (D,M,S; zero is North) ..... "`
270 ...and display azimuth	270 `PRINT Q$+OP$`
275 Repeat the above for sunset (or end of twilight)	275 `PRINT : PRINT "Circumstances of sunset:"`
	280 `PRINT "-----------------------" : GOTO 295`
	285 `PRINT : PRINT "Evening twilight:"`
	290 `PRINT "-----------------"`
295 Now display the results . . .	295 `Q$="Universal time (H,M,S) ............ "`
300 . . . they are already in string format	300 `PRINT : PRINT Q$+UD$`
	305 `Q$="...may be in error by up to 4 minutes"`
310 Announce ambiguity (see TIME)	310 `IF ER(5)=1 THEN PRINT Q$`
	315 `Q$="Local civil time (H,M,S) .......... "`
	320 `PRINT Q$+TD$`
325 No azimuths if twilight	325 `IF SW(7)>0 THEN GOTO 345`
330 Convert to D,M,S format with FND and MINSEC...	330 `X=FND(AD) : SW(1)=1 : GOSUB 1000`
	335 `Q$="Azimuth (D,M,S; zero is North) ..... "`
340 ...and display azimuth	340 `PRINT Q$+OP$`
345 Another go?	345 `Q$="Again (Y or N) ................... "`
350 YESNO returns with the answer	350 `PRINT : GOSUB 960`
	355 `IF E=0 THEN STOP`
360 If FL(3) is 0, we have not made any calculations yet	360 `IF FL(3)=0 THEN FL(4)=0 : PRINT : GOTO 50`
370 FL(4)=0 signifies a new place	365 `Q$="Same place (Y or N) .............. "`
	370 `GOSUB 960 : FL(4)=E`
380 Remember to reset all the relevant date flags ...	375 `Q$="Same date (Y or N) ............... "`
	380 `GOSUB 960 : FL(1)=E : FL(3)=E : FL(6)=E`
385 1: JULDAY, 3: TIME, 6: NUTAT, 7: EQECL	385 `FL(7)=E : PRINT : GOTO 50`

```
INCLUDE YESNO, MINSEC, JULDAY, TIME, OBLIQ, NUTAT, EQECL,
        RISET, ANOMALY, SUN, SUNRS
```

Example

```
Sunrise and sunset and twilight
-------------------------------

Geog. longitude (D,M,S; W neg.) .... ? 2,20,0
Geog. latitude (D,M,S; S neg.) ..... ? 48,52,0
Daylight saving (H ahead of zone t)  ? 2
Time zone (hours; West negative) ... ? 0

Calendar date (D,M,Y) .............. ? 20,7,1984

Calculation type:
Sunrise and sunset (Y or N) ........ ? Y

Circumstances of sunrise:
-------------------------

Universal time (H,M,S) .............      4  8 42.17
Local civil time (H,M,S) ...........      6  8 42.17
Azimuth (D,M,S; zero is North) .....    + 56 27 26.34

Circumstances of sunset:
------------------------

Universal time (H,M,S) .............    19 44 28.17
Local civil time (H,M,S) ...........    21 44 28.17
Azimuth (D,M,S; zero is North) .....    +303 19 57.89

Again (Y or N) ..................... ? Y
Same place (Y or N) ................ ? N
Same date (Y or N) ................. ? N

Geog. longitude (D,M,S; W neg.) .... ? 0,0,0
Geog. latitude (D,M,S; S neg.) ..... ? 52,0,0
Daylight saving (H ahead of zone t)  ? 0
Time zone (hours; West negative) ... ? 0

Calendar date (D,M,Y) .............. ? 1,10,1984

Calculation type:
Sunrise and sunset (Y or N) ........ ? Y

Circumstances of sunrise:
-------------------------

Universal time (H,M,S) .............      6  0 58.12
Local civil time (H,M,S) ...........      6  0 58.12
Azimuth (D,M,S; zero is North) .....    + 94 14 37.49

Circumstances of sunset:
------------------------

Universal time (H,M,S) .............    17 37 10.86
Local civil time (H,M,S) ...........    17 37 10.86
Azimuth (D,M,S; zero is North) .....    +265 27  7.61
Again (Y or N) ..................... ? Y
Same place (Y or N) ................ ? Y
Same date (Y or N) ................. ? N


Calendar date (D,M,Y) .............. ? 4,4,1968
```

3600 SUNRS

```
Calculation type:
Sunrise and sunset (Y or N) ........ ? N
Civil twilight (Y or N) ............ ? N
Nautical twilight (Y or N) ......... ? N
Astronomical twilight (Y or N) ..... ? Y

Morning twilight:
-----------------

Universal time (H,M,S) .............    3 26 17.95
Local civil time (H,M,S) ...........    3 26 17.95

Evening twilight:
-----------------

Universal time (H,M,S) .............   20 41 31.85
Local civil time (H,M,S) ...........   20 41 31.85

Again (Y or N) .................... ? Y
Same place (Y or N) ................ ? N
Same date (Y or N) ................. ? N

Geog. longitude (D,M,S; W neg.) .... ? 149,8,0
Geog. latitude (D,M,S; S neg.) ..... ? -35,18,0
Daylight saving (H ahead of zone t)  ? 1
Time zone (hours; West negative) ... ? 10

Calendar date (D,M,Y) ............. ? 15,11,1984

Calculation type:
Sunrise and sunset (Y or N) ........ ? N
Civil twilight (Y or N) ............ ? N
Nautical twilight (Y or N) ......... ? N
Astronomical twilight (Y or N) ..... ? Y

Morning twilight:
-----------------

Universal time (H,M,S) .............   17 10 24.18
Local civil time (H,M,S) ...........    4 10 24.18

Evening twilight:
-----------------

Universal time (H,M,S) .............   10 26 31.67
Local civil time (H,M,S) ...........   21 26 31.67

Again (Y or N) .................... ? N
```

The Astronomical Almanac gives the following values for local civil times:

Date	Latitude	Longitude	Sunrise	Sunset
20th July 1984	48.87 (Paris)	2.33	06 09	21 45
1st October 1984	52.00 (Cambridge)	0.00	06 01	17 37
		Astronomical twilight:	begin	end
4th April 1968	52.00 (Cambridge)	0.00	03 28	20 42
15th November 1984	−35.30 (Canberra)	149.13	04 10	21 26

3800 PELMENT

This routine finds, at a given instant, the values of the orbital elements of all the major planets, together with their mean daily motions in longitude, angular diameters, and standard visual magnitudes.

The path of any of the major planets in its elliptical orbit around the Sun may be defined in terms of six *orbital elements*: the mean longitude, the longitude of the perihelion, the eccentricity, the inclination, the longitude of the ascending node, and the length of the semi-major axis. Of these, only the last is a true constant. The other elements, being referred to the planes of the ecliptic and the equator, reflect the slow changes caused by the influences of other members of the Solar System. It is therefore necessary to calculate the elements for a given date, and the orbit so defined will then be the correct mean orbit of the planet at that time. The true orbit may be further perturbed by short- and long-period terms which must be considered if high accuracy is required (see sub-routine PLANS, 4500).

The orbital elements are expressed by polynomials of the form

$$A0 + A1T + A2T^2 + A3T^3,$$

where A0, A1, etc. are constants and T is the number of Julian centuries of 36525 Julian days since 1900 January 0.5. Routine PELMENT calculates the values of the orbital elements for *all* the planets on a given date (DY, MN, YR) and returns them in the array PL(8,9). This array must be declared before calling the routine. PELMENT also returns the mean daily motion in longitude, the angular diameter at a distance of 1 AU from the Earth, and the standard visual magnitude. This last quantity is the visual magnitude of the planet when at a distance of 1 AU from both the Sun and the Earth, and with zero phase angle. The first subscript, I, of PL(I,J) indicates the planet by Mercury = 1, Venus = 2, Mars = 3, Jupiter = 4, Saturn = 5, Uranus = 6, Neptune =7, and Pluto = 8. The second subscript, J, indicates the parameter by mean longitude = 1, daily motion in longitude = 2, longitude of the perihelion = 3, eccentricity = 4, inclination = 5, longitude of the ascending node = 6, semi-major axis = 7, angular diameter at 1 AU = 8, and the standard visual magnitude = 9. Thus the longitude of the perihelion of Jupiter is in PL(4,3), while the inclination of Venus is in PL(2,5).

PELMENT accepts a planet name in the string variable P$ as one of the

3800 PELMENT

calling arguments. It tests P$ for validity, returning with the real variable IP = 0 if invalid, or IP set to the planet number (1–8) if P$ is valid. The array PL is completed for all the planets at the given instant provided that P$ is valid. Thus the routine PLANS (4500) is able to correct the position of one planet for perturbations by the others, using the array PL to find the positions as needed.

The mean longitude increases by 360 degrees for every rotation of the planet about the Sun. In order to preserve accuracy, it is expressed in such a manner that integer rotations are subtracted from the second term of the expression before adding the other terms. Thus

$$PL(I,1) = A0 + (A2 \times T^2) + (A3 \times T^3) + B$$
$$B = 360 \times (AA - INT(AA))$$
$$AA = A1 \times T,$$

where INT may represent either the least-integer or truncated-integer function.

Details of PELMENT

Called by GOSUB 3800.

Calculates orbital data for all the major planets at the instant given by DY (days, including the fraction), MN (months) and YR (years). It also checks a planet name input in P$ for validity (upper case letters only), returning with the planet number IP (1–8) if valid, or IP = 0 if not. The orbital data are *not* calculated if IP = 0. The data are returned in the array PL(I,J) with subscripts as follows:

MERCURY	I = 1	SATURN	I = 5
VENUS	I = 2	URANUS	I = 6
MARS	I = 3	NEPTUNE	I = 7
JUPITER	I = 4	PLUTO	I = 8

mean longitude (degrees)	J = 1
mean daily motion in longitude (degrees)	J = 2
longitude or perihelion (degrees)	J = 3
eccentricity	J = 4
inclination (degrees)	J = 5
longitude of the ascending node (degrees)	J = 6
semi-major axis (AU)	J = 7
angular diameter at 1 AU (arcsecond)	J = 8
standard visual magnitude	J = 9

Elements for PLUTO are osculating elements at 21st January 1984 (see ELOSC (7500) for more details of how to deal with osculating elements).
PL(8,9) must be declared before calling the routine.
Other routine called: JULDAY (1100)

Notes		Code
		3797 REM
		3798 REM Subroutine PELMENT
		3799 REM
3800	Data area: planet names . . .	3800 DATA "MERCURY","VENUS","MARS","JUPITER"
		3805 DATA "SATURN","URANUS","NEPTUNE","PLUTO"
3810	Now orbital data for each of the planets in turn . . .	3810 REM Mercury...
3815	Mean longitude	3815 DATA 178.179078,415.2057519,3.011E-4,0
3820	Longitude of the perihelion	3820 DATA 75.899697,1.5554889,2.947E-4,0
3825	Eccentricity	3825 DATA 2.0561421E-1,2.046E-5,-3E-8,0
3830	Inclination	3830 DATA 7.002881,1.8608E-3,-1.83E-5,0
3835	Longitude of the ascending node	3835 DATA 47.145944,1.1852083,1.739E-4,0
3840	Semi-major axis; angular diameter at 1 AU; standard magnitude	3840 DATA 3.870986E-1,6.74,-0.42
3845	. . . and the same for each of the other planets	3845 REM Venus...
		3850 DATA 342.767053,162.5533664,3.097E-4,0
		3855 DATA 130.163833,1.4080361,-9.764E-4,0
		3860 DATA 6.82069E-3,-4.774E-5,9.1E-8,0
		3865 DATA 3.393631,1.0058E-3,-1E-6,0
		3870 DATA 75.779647,8.9985E-1,4.1E-4,0
		3875 DATA 7.233316E-1,16.92,-4.4
		3880 REM Mars...
		3885 DATA 293.737334,53.17137642,3.107E-4,0
		3890 DATA 3.34218203E2,1.8407584,1.299E-4
		3895 DATA -1.19E-6
		3900 DATA 9.33129E-2,9.2064E-5,-7.7E-8,0
		3905 DATA 1.850333,-6.75E-4,1.26E-5,0
		3910 DATA 48.786442,7.709917E-1,-1.4E-6
		3915 DATA -5.33E-6
		3920 DATA 1.5236883,9.36,-1.52
		3925 REM Jupiter...
		3930 DATA 238.049257,8.434172183,3.347E-4
		3935 DATA -1.65E-6
		3940 DATA 1.2720972E1,1.6099617,1.05627E-3
		3945 DATA -3.43E-6
		3950 DATA 4.833475E-2,1.6418E-4,-4.676E-7
		3955 DATA -1.7E-9
		3960 DATA 1.308736,-5.6961E-3,3.9E-6,0
		3965 DATA 99.443414,1.01053,3.5222E-4,-8.51E-6
		3970 DATA 5.202561,196.74,-9.4
		3975 REM Saturn...
		3980 DATA 266.564377,3.398638567,3.245E-4
		3985 DATA -5.8E-6
		3990 DATA 9.1098214E1,1.9584158,8.2636E-4
		3995 DATA 4.61E-6
		4000 DATA 5.589232E-2,-3.455E-4,-7.28E-7
		4005 DATA 7.4E-10
		4010 DATA 2.492519,-3.9189E-3,-1.549E-5,4E-8
		4015 DATA 112.790414,8.731951E-1,-1.5218E-4
		4020 DATA -5.31E-6
		4025 DATA 9.554747,165.6,-8.88

```
4030 REM        Uranus...

4035 DATA 244.19747,1.194065406,3.16E-4,-6E-7
4040 DATA 1.71548692E2,1.4844328,2.372E-4
4045 DATA -6.1E-7
4050 DATA 4.63444E-2,-2.658E-5,7.7E-8,0
4055 DATA 7.72464E-1,6.253E-4,3.95E-5,0
4060 DATA 73.477111,4.986678E-1,1.3117E-3,0
4065 DATA 19.21814,65.8,-7.19

4070 REM        Neptune...

4075 DATA 84.457994,6.107942056E-1,3.205E-4
4080 DATA -6E-7
4085 DATA 4.6727364E1,1.4245744,3.9082E-4
4090 DATA -6.05E-7
4095 DATA 8.99704E-3,6.33E-6,-2E-9,0
4100 DATA 1.779242,-9.5436E-3,-9.1E-6,0
4105 DATA 130.681389,1.098935,2.4987E-4
4110 DATA -4.718E-6
4115 DATA 30.10957,62.2,-6.87
```

4120 No precise theory for PLUTO ...	`4120 REM        Pluto...(osculating 1984 Jan 21)`
4125 ... therefore use osculating elements ...	`4125 DATA 95.3113544,3.980332167E-1,0,0`
4130 ... which are only accurate for a few months ...	`4130 DATA 224.017,0,0,0`
4135 ... near the epoch (21st January 1984). For other dates ...	`4135 DATA 2.5515E-1,0,0,0`
4140 ... look up osculating elements in *The Astronomical Almanac* ...	`4140 DATA 17.1329,0,0,0`
4145 ... and substitute here, or use ELOSC (7500) and HELOSC	`4145 DATA 110.191,0,0,0` `4150 DATA 39.8151,8.2,-1.0`
4170 Reset data pointer to start of DATA area	`4170 RESTORE : IP=0`
4175 Call JULDAY for DJ, and hence calculate T	`4175 GOSUB 1100 : T=(DJ/36525.0)+(UT/8.766E5)`
4180 For each of the planets, compare its name ...	`4180 FOR J=1 TO 8 : READ NM$`
4185 ... with the input name; set IP if a match	`4185 IF NM$=P$ THEN IP=J` `4190 NEXT J`
4195 No match? Error return	`4195 IF IP=0 THEN RETURN`
4200 For each planet ...	`4200 FOR I=1 TO 8`
4205 ... read mean longitude data ...	`4205 READ A0,A1,A2,A3 : AA=A1*T`
4210 ... and calculate first mean longitude at the given ...	`4210 B=360.0*(AA-INT(AA))`
4215 ... epoch (degrees) in range 0–360, and then ...	`4215 C=A0+B+(A3*T+A2)*T*T : GOSUB 4250 : PL(I,1)=C`
4220 ... calculate its mean daily motion in degrees ...	`4220 PL(I,2)=(A1*9.856263E-3)+(A2+A3)/36525.0`
4225 ... read the reset of the data, and place results ...	`4225 FOR J=3 TO 6 : READ A0,A1,A2,A3`

4230 ... in the array PL(8,9)	`4230 PL(I,J)=((A3*T+A2)*T+A1)*T+A0`
	`4235 NEXT J`
	`4240 READ PL(I,7),PL(I,8),PL(I,9)`

4245 ... end of planet loop, and return from PELMENT	`4245 NEXT I : RETURN`

4250 This routine returns C in the range 0–360	`4250 IF C>360.0 THEN C=C-360.0 : GOTO 4250`
	`4255 IF C<0 THEN C=C+360.0 : GOTO 4255`
	`4260 RETURN`

Notes		Code
	1	`REM`
	2	`REM       Handling program HPELMENT`
	3	`REM`
5 Declare array PL	5	`DIM FL(20),ER(20),SW(20),PL(8,9)`
10 FNI is needed by JULDAY	10	`DEF FNI(W)=SGN(W)*INT(ABS(W))`
	15	`PRINT : PRINT`
	20	`PRINT "Planetary elements: mean equinox of date"`
	25	`PRINT "----------------------------------------"`
	30	`PRINT : PRINT`
35 FL(1)=0 for new date	35	`IF FL(1)=1 THEN GOTO 50`
40 Get a new date . . .	40	`Q$="Greenwich date (D,M,Y) ..................... "`
	45	`PRINT Q$; : INPUT DY,MN,YR`
50 . . . and a planet name	50	`Q$="Which planet (full name, upper case) ....... "`
	55	`PRINT Q$;  : INPUT P$`
60 Call PELMENT	60	`UT=0 : GOSUB 3800`
65 Was the date impossible . . .	65	`IF ER(1)=1 THEN GOTO 185`
70 . . . or the name of the planet unrecognised?	70	`IF IP=0 THEN PRINT "What?" : GOTO 50`
75 Display the results for planet P$ . . .	75	`PRINT : PRINT "Planet "; P$; ":" : PRINT`
80 . . . if PLUTO print a warning . . .	80	`Q$="Osculating elements at 1984 Jan 21"`
	85	`IF IP=8 THEN PRINT Q$`
95 . . . in the format D,M,S via MINSEC	90	`Q$="Mean longitude (D,M,S) ..................... "`
	95	`SW(1)=1 : X=PL(IP,1) : GOSUB 1000`
	100	`PRINT Q$+OP$`
	105	`Q$="Daily motion in mean long (D,M,S) .......... "`
	110	`X=PL(IP,2) : GOSUB 1000 : PRINT Q$+OP$`
	115	`Q$="Longitude of perihelion (D,M,S) ............ "`
	120	`X=PL(IP,3) : GOSUB 1000 : PRINT Q$+OP$`
	125	`Q$="Eccentricity ............................... "`
130 Round to six decimal places	130	`PRINT Q$+" "; INT((PL(IP,4)*1E6)+0.5)/1E6`
	135	`Q$="Inclination (D,M,S) ........................ "`
	140	`X=PL(IP,5) : GOSUB 1000 : PRINT Q$+OP$`
	145	`Q$="Longitude of ascending node (D,M,S) ........ "`
	150	`X=PL(IP,6) : GOSUB 1000 : PRINT Q$+OP$`
	155	`Q$="Length of semi-major axis (AU) ............. "`
160 Round to six decimal places	160	`PRINT Q$+" "; INT((PL(IP,7)*1E6)+0.5)/1E6`
	165	`Q$="Angular diameter at 1 AU (D,M,S) .......... "`
	170	`X=PL(IP,8)/3600.0 : GOSUB 1000 : PRINT Q$+OP$`
	175	`Q$="Visual magnitude V(1,0) .................... "`
	180	`PRINT Q$+" "; PL(IP,9)`
185 Another go?	185	`Q$="Again (Y or N) ............................. "`
190 Call YESNO for an answer	190	`PRINT : GOSUB 960`
	195	`IF E=0 THEN STOP`
200 If FL(1)=0 we have not yet done anything	200	`IF FL(1)=0 THEN PRINT : GOTO 35`
	205	`Q$="Same date (Y or N) ......................... "`
210 Set FL(1)=0 for a new date	210	`GOSUB 960 : FL(1)=E : PRINT : GOTO 35`

```
INCLUDE YESNO, MINSEC, JULDAY, PELMENT
```

3800 PELMENT

Example

```
Planetary elements: mean equinox of date
----------------------------------------


Greenwich date (D,M,Y) ..................... ? 24,7,1992
Which planet (full name, upper case) ....... ? JUPITER

Planet JUPITER:

Mean longitude (D,M,S) ..................... +168 26 50.41
Daily motion in mean long (D,M,S) .......... +  0  4 59.27
Longitude of perihelion (D,M,S) ............ + 14 12 43.39
Eccentricity ...............................   0.048486
Inclination (D,M,S) ........................ +  1 18 12.48
Longitude of ascending node (D,M,S) ........ +100 22 44.60
Length of semi-major axis (AU) .............   5.202561
Angular diameter at 1 AU (D,M,S) ........... +  0  3 16.74
Visual magnitude V(1,0) ....................   -9.4

Again (Y or N) ............................. ? Y
Same date (Y or N) ......................... ? N

Greenwich date (D,M,Y) ..................... ? 21,1,1984
Which planet (full name, upper case) ....... ? MERCURY

Planet MERCURY:

Mean longitude (D,M,S) ..................... +176 12  0.06
Daily motion in mean long (D,M,S) .......... +  4  5 32.56
Longitude of perihelion (D,M,S) ............ + 77 12 26.45
Eccentricity ...............................   0.205631
Inclination (D,M,S) ........................ +  7  0 15.96
Longitude of ascending node (D,M,S) ........ + 48  8 32.19
Length of semi-major axis (AU) .............   0.387099
Angular diameter at 1 AU (D,M,S) ........... +  0  0  6.74
Visual magnitude V(1,0) ....................   -0.42

Again (Y or N) ............................. ? N
```

4500 PLANS

This routine calculates the instantaneous heliocentric ecliptic coordinates, radius vector and distance from the Earth, and the apparent geocentric ecliptic coordinates (allowing for light-travel time), of any of the major planets at a given instant.

The geocentric position of any of the eight major planets can be calculated from its orbital elements returned by the subroutine PELMENT (3800). These quantities define the mean orbit of the planet at the given instant, and several refinements must be made to find, with greater precision, the apparent position as seen from the Earth. The first refinement is to make allowances for the perturbations to the mean orbit from other members of the Solar System. These are complex and depend upon the relative positions of the planets in their separate orbits around the Sun. They are best expressed as series of terms like

$$p \sin (\omega t) + q \cos (\omega t),$$

where the amplitudes p and q are chosen to fit the observed departures from the mean orbit at angular frequency ω and time t. Many such terms are needed, the number generally increasing with a planet's distance from the Sun. Thus, to achieve a given accuracy, Mercury needs 4 perturbation terms in the longitude, while Saturn requires 24 in the mean longitude, 49 in the eccentricity, 17 in the longitude of the perihelion, 36 in the length of the semi-major axis, and finally 6 in the heliocentric latitude. The routine given here takes account of perturbations for all the planets except Pluto. No dynamical theory exists for this outermost of the planets and its orbit is usually defined in terms of constantly-changing *osculating elements*. PLANS uses the osculating elements for 21st January 1984. Calculations will be quite precise near to that date, but will become progressively less accurate as the time interval between then and the required date increases. You can make precise calculations for PLUTO near any date provided that you use the correct osculating elements, substituted either in place of those given in PELMENT, or used directly in routine ELOSC (7500) and its handling program. Osculating elements for many members of the Solar System can be found in the yearly publication *The Astronomical Almanac* which ought to be available from your local library.

The second refinement is to take account of the finite light-travel time between the Earth and the planet in question. When we view a planet now, we see it in the position it occupied t hours ago, given by

$t = 0.1386 \times RH,$

where RH is the distance in AU between the Earth and the planet. In this routine, an approximate position for the planet is first calculated, neglecting the light-travel time. Then a second pass is made through the program using the light-travel time based on the approximate position found on the first pass.

Further refinements include making allowances for nutation by adding DP to the geocentric longitude and DO to the obliquity of the ecliptic (see routine NUTAT, 1800), and taking account of aberration, the apparent displacement in the position due to the Earth's orbital velocity (see routine SUN, 3400). A planet with geocentric ecliptic longitude EP and latitude BP has an apparent position EP1 and BP1 given (in *radians*) by

$EP1 = EP - (9.9387 \times 10^{-5} \times COS(A)/COS(BP))$

$BP1 = BP - (9.9387 \times 10^{-5} \times SIN(A) \times SIN(BP)),$

where $A = LG + PI - EP$, $PI = 3.1415927$, and LG is the longitude of the Earth in its own orbit round the Sun. When the observed body *is* the Sun, we have $EP = LG + PI$ so that $A = 0$. Thus, for the Sun, $EP1 = EP - 9.9387 \times 10^{-5}$ (since $BP = 0$) which is precisely the term used in the handling program of SUN (3400), having made the appropriate conversion to degrees.

Subroutine PLANS accepts a given instant as DY (days including the fraction), MN (months), YR (years), and a planet's name in the string variable P$ (upper case, full name, e.g. MERCURY). It returns with the instantaneous heliocentric ecliptic longitude, L0, latitude, S0, radius vector, P0, and distance from the Earth, V0, all corrected for perturbations but *not* for light-travel time. These quantities are therefore the true values at the given instant. PLANS also calculates the geocentric ecliptic longitude, EP, latitude, BP, and distance from the Earth, RH, corrected both for perturbations and light-travel time. These are then the *apparent* values as seen from the centre of the Earth at the given instant. Corrections for nutation and aberration must be made outside the subroutine, in this case in lines 200–215 of the handling program, HPLANS. The right ascension and declination calculated from the fully-corrected ecliptic coordinates (see EQECL, 2000) are then the apparent geocentric coordinates. Further corrections can be made, if required, for atmospheric refraction (REFRACT, 3000) and geocentric parallax (PARALLX, 2800), although the intrinsic error in PLANS of about 10 arcseconds is usually the dominant error at this stage of the calculation.

Other routines called by PLANS include PELMENT (3800), SUN (3400), and ANOMALY (3300). Other calculations made by the handling program include finding the solar elongation, E, the apparent angular diameter, TH, the phase or fraction illuminated, F (full = 1), and the approximate visual magnitude, M. The formulae used are listed below. The arrays PL(8,9) and AP(8) must both be declared before calling the routine.

Formulae

For an inner planet:

$TAN(EP-LG-PI) = -RD \times SIN(PD-LG)/(RE-(RD \times COS(PD-LG)))$

For an outer planet:

$TAN(EP-PD) = RE \times SIN(PD-LG)/(RD-(RE \times COS(PD-LG)))$

For inner and outer alike:

$TAN(BP) = RD \times SIN(PS) \times SIN(EP-PD)/(COS(PS) \times RE \times SIN(PD-LG))$

$RH = RE^2 + PV^2 - (2 \times RE \times PV \times COS(PS) \times COS(PD-LG))$

$LI = RH \times 5.775518E-3$

$RD = PV \times COS(PS)$

$TAN(PD-OM) = SIN(LP-OM) \times COS(IN)/COS(LP-OM)$

$SIN(PS) = SIN(LP-OM) \times SIN(IN)$

$PV = A \times (1-EC^2)/(1+(EC \times COS(AT)))$

$LP = AT + PL(IP,3)$

$AM = L - PL(IP,3) - (LI \times PL(IP,2))$

AT = true anomaly from AM via Kepler's equation

L = mean longitude

PL(IP,3) = longitude of the perihelion

PL(IP,2) = mean daily motion in longitude

EC = eccentricity

A = semi-major axis

OM = longitude of the ascending node

RE = RR = radius of Earth's orbit (SUN, 3300)

LG = SR+PI = longitude of Earth (SUN, 3300)

PI = 3.1415927

IN = inclination

IP = planet number (1–8)

Perturbation corrections are applied to some or all of longitude (QA), radius of planet's orbit (QB), mean longitude (QC), eccentricity (QD), mean anomaly (QE), semi-major axis (QF), heliocentric longitude (QG), and the longitude of the perihelion (VK), depending on the planet.

$COS(E) = -COS(BP) \times COS(EP-LG)$

$TH = PL(IP,8)/RH$

PL(IP,8) = angular diameter at 1 AU distance

$F = 0.5 \times (1+COS(EP-PD))$

$M = 5 \times LOG_{10}(PV \times RH/F^{0.5}) + PL(IP,9)$

PL(IP,9) = standard magnitude.

4500 PLANS

Details of PLANS

Called by GOSUB 4500.

For planet P$ (full name, upper case), on date DY (days, including the fraction), MN (months), YR (years) calculates: L0, the heliocentric longitude, S0, the heliocentric latitude, P0, the radius vector of the planet, and V0, the distance from the Earth, not corrected for light-travel time. Also calculates EP, the geocentric ecliptic longitude, BP, the geocentric ecliptic latitude, and RH, the Earth–planet distance, corrected for light-travel time. All angles in *radians* and distances in *AU*.

Arrays PL(8,9) and AP(8) must be declared before calling the routine.

Other routines called: PELMENT (3800), ANOMALY (3300), SUN (2400).

	4497 REM
	4498 REM Subroutine PLANS
	4499 REM
4500 Call PELMENT to fill PL(8,9)	4500 GOSUB 3800 : LI=0 : PI=3.1415926536 : TP=2.0*PI
4505 Return if error, either planet name unrecognised . . .	4505 IF IP=0 THEN RETURN
4510 . . . or impossible date (JULDAY)	4510 IF ER(1)=1 THEN RETURN
4515 Call SUN and set Earth longitude (LG) and radius vector (RE)	4515 GOSUB 3400 : MS=AM : RE=RR : LG=SR+PI
4520 Two passes, the second to take care of light-travel time	4520 FOR K=1 TO 2
4525 Set up mean anomalies of all planets in AP(8)	4525 FOR J=1 TO 8
	4530 AP(J)=FNM(PL(J,1)-PL(J,3)-LI*PL(J,2))
	4535 NEXT J
4540 Initialise perturbation terms to 0 . . .	4540 QA=0 : QB=0 : QC=0 : QD=0 : QE=0 : QF=0 : QG=0
4545 . . . and then calculate those that are needed . . .	4545 IF IP<5 THEN ON IP GOSUB 4685,4735,4810,4945
4550 . . . by calling up appropriate local routine	4550 IF IP>4 THEN ON IP-4 GOSUB 4945,4945,4945,5760
4560 Set eccentricity, mean anomaly, and call ANOMALY	4560 EC=PL(IP,4)+QD : AM=AP(IP)+QE : GOSUB 3300
4565 PV is radius vector of planet	4565 PV=(PL(IP,7)+QF)*(1.0-EC*EC)/(1.0+EC*COS(AT))
4570 LP is planet's orbital longitude	4570 LP=FND(AT)+PL(IP,3)+FND(QC-QE) : LP=FNM(LP)
4575 OM is longitude of ascending node	4575 OM=FNM(PL(IP,6)) : LO=LP-OM
4580 Calculate trigonometric values only once	4580 SO=SIN(LO) : CO=COS(LO)
4585 IN is the inclination	4585 IN=FNM(PL(IP,5)) : PV=PV+QB
	4590 SP=SO*SIN(IN) : Y=SO*COS(IN)
4595 PS is the heliocentric ecliptic latitude	4595 PS=FNS(SP)+QG : SP=SIN(PS)
	4600 PD=ATN(Y/CO)+OM+FNM(QA)
4605 Correct for ambiguity of inverse tangent . . .	4605 IF CO<0 THEN PD=PD+PI
4610 . . . and restore to primary interval (0–2π)	4610 IF PD>TP THEN PD=PD-TP
	4615 CI=COS(PS) : RD=PV*CI : LL=PD-LG
	4620 RH=RE*RE+PV*PV-2.0*RE*PV*CI*COS(LL)
4625 RH is Earth–planet distance; LI is light-travel time (days)	4625 RH=SQR(RH) : LI=RH*5.775518E-3
4630 Catch true values on first pass (not corrected for LI)	4630 IF K=1 THEN LO=PD : VO=RH : SO=PS : PO=PV
4635 End of light-travel time loop	4635 NEXT K
4640 Calculate trigonometric values just once	4640 L1=SIN(LL) : L2=COS(LL)
4645 Inner planet or outer?	4645 IF IP<3 THEN GOTO 4655
4650 Outer: calculate geocentric ecliptic longitude (EP) . . .	4650 EP=ATN(RE*L1/(RD-RE*L2))+PD : GOTO 4660
4655 Inner: calculate geocentric ecliptic longitude (EP) . . .	4655 EP=ATN(-1.0*RD*L1/(RE-RD*L2))+LG+PI
4660 . . . and restore to primary interval	4660 IF EP<0 THEN EP=EP+TP : GOTO 4660
	4665 IF EP>TP THEN EP=EP-TP : GOTO 4665
4670 Calculate geocentric ecliptic latitude (BP)	4670 BP=ATN(RD*SP*SIN(EP-PD)/(CI*RE*L1))
4675 Normal return from PLANS	4675 RETURN
4680 Here follow local perturbation routines; first Mercury	4680 REM Perturbations for Mercury
4685 QA: perturbation in longitude	4685 QA=2.04E-3*COS(5*AP(2)-2*AP(1)+2.1328E-1)
	4690 QA=QA+1.03E-3*COS(2*AP(2)-AP(1)-2.8046)
	4695 QA=QA+9.1E-4*COS(2*AP(4)-AP(1)-6.4582E-1)
	4700 QA=QA+7.8E-4*COS(5*AP(2)-3*AP(1)+1.7692E-1)
4705 QB: perturbation in radius vector	4705 QB=7.525E-6*COS(2*AP(4)-AP(1)+9.25251E-1)
	4710 QB=QB+6.802E-6*COS(5*AP(2)-3*AP(1)-4.53642)
	4715 QB=QB+5.457E-6*COS(2*AP(2)-2*AP(1)-1.24246)

4500 PLANS

4725   That's all for Mercury

4730   Perturbations for Venus

4735   QC: perturbation in mean longitude

4740   QA: perturbation in longitude

4765   QB: perturbation in radius vector

4800   That's all for Venus

4805   Perturbations for Mars

4815   QC: perturbation in mean longitude

4820   QA: perturbation in longitude

4865   QB: perturbation in radius vector

4930   That's all for Mars

4935   Jupiter, Saturn, Uranus, Neptune share this bit . . .

4945   FNU is an unwinding function (see HPLANS)

4965   . . . end of common code; now jump to the appropriate . . .

4970   . . . perturbation routine . . .

4980   . . . except Jupiter and Saturn share another bit of code . . .

```
4720 QB=QB+3.569E-6*COS(5*AP(2)-AP(1)-1.35699)
4725 RETURN

4730 REM      ....Venus

4735 QC=7.7E-4*SIN(4.1406+T*2.6227) : QC=FNM(QC) : QE=QC

4740 QA=3.13E-3*COS(2*MS-2*AP(2)-2.587)
4745 QA=QA+1.98E-3*COS(3*MS-3*AP(2)+4.4768E-2)
4750 QA=QA+1.36E-3*COS(MS-AP(2)-2.0788)
4755 QA=QA+9.6E-4*COS(3*MS-2*AP(2)-2.3721)
4760 QA=QA+8.2E-4*COS(AP(4)-AP(2)-3.6318)

4765 QB=2.2501E-5*COS(2*MS-2*AP(2)-1.01592)
4770 QB=QB+1.9045E-5*COS(3*MS-3*AP(2)+1.61577)
4775 QB=QB+6.887E-6*COS(AP(4)-AP(2)-2.06106)
4780 QB=QB+5.172E-6*COS(MS-AP(2)-5.08065E-1)
4785 QB=QB+3.62E-6*COS(5*MS-4*AP(2)-1.81877)
4790 QB=QB+3.283E-6*COS(4*MS-4*AP(2)+1.10851)
4795 QB=QB+3.074E-6*COS(2*AP(4)-2*AP(2)-9.62846E-1)
4800 RETURN

4805 REM      ....Mars

4810 A=3*AP(4)-8*AP(3)+4*MS : SA=SIN(A) : CA=COS(A)
4815 QC=-(1.133E-2*SA+9.33E-3*CA) : QC=FNM(QC) : QE=QC

4820 QA=7.05E-3*COS(AP(4)-AP(3)-8.5448E-1)
4825 QA=QA+6.07E-3*COS(2*AP(4)-AP(3)-3.2873)
4830 QA=QA+4.45E-3*COS(2*AP(4)-2*AP(3)-3.3492)
4835 QA=QA+3.88E-3*COS(MS-2*AP(3)+3.5771E-1)
4840 QA=QA+2.38E-3*COS(MS-AP(3)+6.1256E-1)
4845 QA=QA+2.04E-3*COS(2*MS-3*AP(3)+2.7688)
4850 QA=QA+1.77E-3*COS(3*AP(3)-AP(2)-1.0053)
4855 QA=QA+1.36E-3*COS(2*MS-4*AP(3)+2.6894)
4860 QA=QA+1.04E-3*COS(AP(4)+3.0749E-1)

4865 QB=5.3227E-5*COS(AP(4)-AP(3)+7.17864E-1)
4870 QB=QB+5.0989E-5*COS(2*AP(4)-2*AP(3)-1.77997)
4875 QB=QB+3.8278E-5*COS(2*AP(4)-AP(3)-1.71617)
4880 QB=QB+1.5996E-5*COS(MS-AP(3)-9.69618E-1)
4885 QB=QB+1.4764E-5*COS(2*MS-3*AP(3)+1.19768)
4890 QB=QB+8.966E-6*COS(AP(4)-2*AP(3)+7.61225E-1)
4895 QB=QB+7.914E-6*COS(3*AP(4)-2*AP(3)-2.43887)
4900 QB=QB+7.004E-6*COS(2*AP(4)-3*AP(3)-1.79573)
4905 QB=QB+6.62E-6*COS(MS-2*AP(3)+1.97575)
4910 QB=QB+4.93E-6*COS(3*AP(4)-3*AP(3)-1.33069)
4915 QB=QB+4.693E-6*COS(3*MS-5*AP(3)+3.32665)
4920 QB=QB+4.571E-6*COS(2*MS-4*AP(3)+4.27086)
4925 QB=QB+4.409E-6*COS(3*AP(4)-AP(3)-2.02158)
4930 RETURN

4935 REM      ....Jupiter, Saturn
4940 REM      ....Uranus, and Neptune
4945 J1=T/5.0+0.1 : J2=FNU(4.14473+5.29691E1*T)
4950 J3=FNU(4.641118+2.132991E1*T)
4955 J4=FNU(4.250177+7.478172*T)
4960 J5=5.0*J3-2.0*J2 : J6=2.0*J2-6.0*J3+3.0*J4

4965 IF IP<5 THEN ON IP GOTO 5190,5190,5190,4980

4970 IF IP>4 THEN ON IP-4 GOTO 4980,5500,5500,5190

4980 J7=J3-J2 : U1=SIN(J3) : U2=COS(J3) : U3=SIN(2.0*J3)
4985 U4=COS(2.0*J3) : U5=SIN(J5) : U6=COS(J5)
4990 U7=SIN(2.0*J5) : U8=SIN(J6) : U9=SIN(J7)
```

       ☛

```
4995 UA=COS(J7) : UB=SIN(2.0*J7) : UC=COS(2.0*J7)
5000 UD=SIN(3.0*J7) : UE=COS(3.0*J7) : UF=SIN(4.0*J7)
5005 UG=COS(4.0*J7) : VH=COS(5.0*J7)
```

5010 ...end of common code; jump to local routine if Saturn

```
5010 IF IP=5 THEN GOTO 5200
```

5015 Perturbation routine for Jupiter

```
5015 REM     ....Jupiter
```

5020 QC: perturbation in mean longitude

```
5020 QC=(3.31364E-1-(1.0281E-2+4.692E-3*J1)*J1)*U5
5025 QC=QC+(3.228E-3-(6.4436E-2-2.075E-3*J1)*J1)*U6
5030 QC=QC-(3.083E-3+(2.75E-4-4.89E-4*J1)*J1)*U7
5035 QC=QC+2.472E-3*U8+1.3619E-2*U9+1.8472E-2*UB
5040 QC=QC+6.717E-3*UD+2.775E-3*UF+6.417E-3*UB*U1
5045 QC=QC+(7.275E-3-1.253E-3*J1)*U9*U1+2.439E-3*UD*U1
5050 QC=QC-(3.5681E-2+1.208E-3*J1)*U9*U2-3.767E-3*UC*U1
5055 QC=QC-(3.3839E-2+1.125E-3*J1)*UA*U1-4.261E-3*UB*U2
5060 QC=QC+(1.161E-3*J1-6.333E-3)*UA*U2+2.178E-3*U2
5065 QC=QC-6.675E-3*UC*U2-2.664E-3*UE*U2-2.572E-3*U9*U3
5070 QC=QC-3.567E-3*UB*U3+2.094E-3*UA*U4+3.342E-3*UC*U4
5075 QC=FNM(QC)
```

5080 QD: perturbation in eccentricity

```
5080 QD=(3606+(130-43*J1)*J1)*U5+(1289-580*J1)*U6
5085 QD=QD-6764*U9*U1-1110*UB*U1-224*UD*U1-204*U1
5090 QD=QD+(1284+116*J1)*UA*U1+188*UC*U1
5095 QD=QD+(1460+130*J1)*U9*U2+224*UB*U2-817*U2
5100 QD=QD+6074*U2*UA+992*UC*U2+508*UE*U2+230*UG*U2
5105 QD=QD+108*VH*U2-(956+73*J1)*U9*U3+448*UB*U3
5110 QD=QD+137*UD*U3+('108*J1-997)*UA*U3+480*UC*U3
5115 QD=QD+148*UE*U3+(99*J1-956)*U9*U4+490*UB*U4
5120 QD=QD+158*UD*U4+179*U4+(1024+75*J1)*UA*U4
5125 QD=QD-437*UC*U4-132*UE*U4 : QD=QD*1E-7
```

5130 VK: perturbation in longitude of perihelion

```
5130 VK=(7.192E-3-3.147E-3*J1)*U5-4.344E-3*U1
5135 VK=VK+(J1*(1.97E-4*J1-6.75E-4)-2.0428E-2)*U6
5140 VK=VK+3.4036E-2*UA*U1+(7.269E-3+6.72E-4*J1)*U9*U1
5145 VK=VK+5.614E-3*UC*U1+2.964E-3*UE*U1+3.7761E-2*U9*U2
5150 VK=VK+6.158E-3*UB*U2-6.603E-3*UA*U2-5.356E-3*U9*U3
5155 VK=VK+2.722E-3*UB*U3+4.483E-3*UA*U3
5160 VK=VK-2.642E-3*UC*U3+4.403E-3*U9*U4
5165 VK=VK-2.536E-3*UB*U4+5.547E-3*UA*U4-2.689E-3*UC*U4
```

5170 QE: perturbation in mean anomaly

```
5170 QE=QC-(FNM(VK)/PL(IP,4))
```

5175 QF: perturbation in semi-major axis

```
5175 QF=205*UA-263*U6+693*UC+312*UE+147*UG+299*U9*U1
5180 QF=QF+181*UC*U1+204*UB*U2+111*UD*U2-337*UA*U2
5185 QF=QF-111*UC*U2 : QF=QF*1E-6
```

5190 That's all for Jupiter

```
5190 RETURN
```

5195 Perturbations for Saturn

```
5195 REM     ....Saturn
```

5200 First work out some arguments

```
5200 UI=SIN(3*J3) : UJ=COS(3*J3) : UK=SIN(4*J3)
5205 UL=COS(4*J3) : VI=COS(2*J5) : UN=SIN(5*J7)
5210 J8=J4-J3 : UO=SIN(2*J8) : UP=COS(2*J8)
5215 UQ=SIN(3*J8) : UR=COS(3*J8)
```

5220 QC: perturbation in mean longitude

```
5220 QC=7.581E-3*U7-7.986E-3*U8-1.48811E-1*U9
5225 QC=QC-(8.14181E-1-(1.815E-2-1.6714E-2*J1)*J1)*U5
5230 QC=QC-(1.0497E-2-(1.60906E-1-4.1E-3*J1)*J1)*U6
5235 QC=QC-1.5208E-2*UD-6.339E-3*UF-6.244E-3*U1
5240 QC=QC-1.65E-2*UB*U1-4.0786E-2*UB
5245 QC=QC+(8.931E-3+2.728E-3*J1)*U9*U1-5.775E-3*UD*U1
5250 QC=QC+(8.1344E-2+3.206E-3*J1)*UA*U1+1.5019E-2*UC*U1
5255 QC=QC+(8.5581E-2+2.494E-3*J1)*U9*U2+1.4394E-2*UC*U2
5260 QC=QC+(2.5328E-2-3.117E-3*J1)*UA*U2+6.319E-3*UE*U2
5265 QC=QC+6.369E-3*U9*U3+9.156E-3*UB*U3+7.525E-3*UQ*U3
5270 QC=QC-5.236E-3*UA*U4-7.736E-3*UC*U4-7.528E-3*UR*U4
5275 QC=FNM(QC)
```

| 5280 | QD: perturbation in eccentricity | ```
5280  QD=(-7927+(2548+91*J1)*J1)*U5
5285  QD=QD+(13381+(1226-253*J1)*J1)*U6+(248-121*J1)*U7
5290  QD=QD-(305+91*J1)*VI+412*UB+12415*U1
5295  QD=QD+(390-617*J1)*U9*U1+(165-204*J1)*UB*U1
5300  QD=QD+26599*UA*U1-4687*UC*U1-1870*UE*U1-821*UG*U1
5305  QD=QD-377*VH*U1+497*UP*U1+(163-611*J1)*U2
5310  QD=QD-12696*U9*U2-4200*UB*U2-1503*UD*U2-619*UF*U2
5315  QD=QD-268*UN*U2-(282+1306*J1)*UA*U2
5320  QD=QD+(-86+230*J1)*UC*U2+461*UO*U2-350*U3
5325  QD=QD+(2211-286*J1)*U9*U3-2208*UB*U3-568*UD*U3
5330  QD=QD-346*UF*U3-(2780+222*J1)*UA*U3
5335  QD=QD+(2022+263*J1)*UC*U3+248*UE*U3+242*UQ*U3
5340  QD=QD+467*UR*U3-490*U4-(2842+279*J1)*U9*U4
5345  QD=QD+(128+226*J1)*UB*U4+224*UD*U4
5350  QD=QD+(-1594+282*J1)*UA*U4+(2162-207*J1)*UC*U4
5355  QD=QD+561*UE*U4+343*UG*U4+469*UQ*U4-242*UR*U4
5360  QD=QD-205*U9*UI+262*UD*UI+208*UA*UJ-271*UE*UJ
5365  QD=QD-382*UE*UK-376*UD*UL : QD=QD*1E-7
``` |
| 5370 | VK: perturbation in longitude of peri-<br>helion | ```
5370  VK=(7.7108E-2+(7.186E-3-1.533E-3*J1)*J1)*U5
5375  VK=VK-7.075E-3*U9
5380  VK=VK+(4.5803E-2-(1.4766E-2+5.36E-4*J1)*J1)*U6
5385  VK=VK-7.2586E-2*U2-7.5825E-2*U9*U1-2.4839E-2*UB*U1
5390  VK=VK-8.631E-3*UD*U1-1.50383E-1*UA*U2
5395  VK=VK+2.6897E-2*UC*U2+1.0053E-2*UE*U2
5400  VK=VK-(1.3597E-2+1.719E-3*J1)*U9*U3+1.1981E-2*UB*U4
5405  VK=VK-(7.742E-3-1.517E-3*J1)*UA*U3
5410  VK=VK+(1.3586E-2-1.375E-3*J1)*UC*U3
5415  VK=VK-(1.3667E-2-1.239E-3*J1)*U9*U4
5420  VK=VK+(1.4861E-2+1.136E-3*J1)*UA*U4
5425  VK=VK-(1.3064E-2+1.628E-3*J1)*UC*U4
``` |
| 5430 | QE: perturbation in mean anomaly | ```
5430  QE=QC-(FNM(VK)/PL(IP,4))
``` |
| 5435 | QF: perturbation in semi-major axis | ```
5435  QF=572*U5-1590*UB*U2+2933*U6-647*UD*U2
5440  QF=QF+33629*UA-344*UF*U2-3081*UC+2885*UA*U2
5445  QF=QF-1423*UE+(2172+102*J1)*UC*U2-671*UG
5450  QF=QF+296*UE*U2-320*VH-267*UB*U3+1098*U1
5455  QF=QF-778*UA*U3-2812*U9*U1+495*UC*U3+688*UB*U1
5460  QF=QF+250*UE*U3-393*UD*U1-856*U9*U4-228*UF*U1
5465  QF=QF+441*UB*U4+2138*UA*U1+296*UC*U4-999*UC*U1
5470  QF=QF+211*UE*U4-642*UE*U1-427*U9*UI-325*UG*U1
5475  QF=QF+398*UD*UI-890*U2+344*UA*UJ+2206*U9*U2
5480  QF=QF-427*UE*UJ : QF=QF*1E-6
``` |
| 5485 | QB: perturbation in heliocentric<br>longitude | ```
5485  QG=7.47E-4*UA*U1+1.069E-3*UA*U2+2.108E-3*UB*U3
5490  QG=QG+1.261E-3*UC*U3+1.236E-3*UB*U4-2.075E-3*UC*U4
``` |
| 5495 | That's all for Saturn | ```
5495  QG=FNM(QG) : RETURN
``` |
| 5500 | Perturbations for Uranus and<br>Neptune . . . | ```
5500  REM     ....Uranus, and Neptune
``` |
| 5505 | . . . share some common code | ```
5505  J8=FNU(1.46205+3.81337*T) : J9=2*J8-J4
5510  VJ=SIN(J9) : UU=COS(J9) : UV=SIN(2*J9)
5515  UW=COS(2*J9)
``` |
| 5520 | . . . end of common code; jump if<br>Neptune | ```
5520  IF IP=7 THEN GOTO 5670
``` |
| 5525 | Perturbations for Uranus | ```
5525  REM     ....Uranus
``` |
| 5530 | Work out some arguments | ```
5530  JA=J4-J2 : JB=J4-J3 : JC=J8-J4
``` |
| 5535 | QC: perturbation in mean longitude | ```
5535  QC=(8.64319E-1-1.583E-3*J1)*VJ
5540  QC=QC+(8.2222E-2-6.833E-3*J1)*UU+3.6017E-2*UV
5545  QC=QC-3.019E-3*UW+8.122E-3*SIN(J6) : QC=FNM(QC)
``` |
| 5550 | VK: perturbation in longitude of peri-<br>helion | ```
5550  VK=1.20303E-1*VJ+6.197E-3*UV
5555  VK=VK+(1.9472E-2-9.47E-4*J1)*UU
``` |
| 5560 | QE: perturbation in mean anomaly | ```
5560  QE=QC-(FNM(VK)/PL(IP,4))
``` |
| 5565 | QD: perturbation in eccentricity | ```
5565  QD=(163*J1-3349)*VJ+20981*UU+1311*UW : QD=QD*1E-7
``` |
```

| 5570 | QF: perturbation in semi-major axis |

```
5570  QF=-3.825E-3*UU
```

| 5580 | QA: perturbation in longitude |

```
5580  QA=(-3.8581E-2+(2.031E-3-1.91E-3*J1)*J1)*COS(J4+JB)
5585  QA=QA+(1.0122E-2-9.88E-4*J1)*SIN(J4+JB)
5590  A=(3.4964E-2-(1.038E-3-8.68E-4*J1)*J1)*COS(2*J4+JB)
5595  QA=A+QA+5.594E-3*SIN(J4+3*JC)-1.4808E-2*SIN(JA)
5600  QA=QA-5.794E-3*SIN(JB)+2.347E-3*COS(JB)
5605  QA=QA+9.872E-3*SIN(JC)+8.803E-3*SIN(2*JC)
5610  QA=QA-4.308E-3*SIN(3*JC)
```

| 5615 | Work out trigonometric values just once |

```
5615  UX=SIN(JB) : UY=COS(JB) : UZ=SIN(J4)
5620  VA=COS(J4) : VB=SIN(2*J4) : VC=COS(2*J4)
```

| 5625 | QG: perturbation in heliocentric longitude |

```
5625  QG=(4.58E-4*UX-6.42E-4*UY-5.17E-4*COS(4*JC))*UZ
5630  QG=QG-(3.47E-4*UX+8.53E-4*UY+5.17E-4*SIN(4*JB))*VA
5635  QG=QG+4.03E-4*(COS(2*JC)*VB+SIN(2*JC)*VC)
5640  QG=FNM(QG)
```

| 5645 | QB: perturbation in radius vector |

```
5645  QB=-25948+4985*COS(JA)-1230*VA+3354*UY
5650  QB=QB+904*COS(2*JC)+894*(COS(JC)-COS(3*JC))
5655  QB=QB+(5795*VA-1165*UZ+1388*VC)*UX
5660  QB=QB+(1351*VA+5702*UZ+1388*VB)*UY
```

| 5665 | That's all for Uranus |

```
5665  QB=QB*1E-6 : RETURN
```

| 5670 | perturbations for Neptune |

```
5670  REM       ....Neptune
```

| 5675 | Arguments ... |
| 5680 | QC: perturbation in mean longitude |

```
5675  JA=J8-J2 : JB=J8-J3 : JC=J8-J4
5680  QC=(1.089E-3*J1-5.89833E-1)*VJ
5685  QC=QC+(4.658E-3*J1-5.6094E-2)*UU-2.4286E-2*UV
5690  QC=FNM(QC)
```

| 5695 | VK: perturbation in longitude of perihelion |

```
5695  VK=2.4039E-2*VJ-2.5303E-2*UU+6.206E-3*UV
```

| 5700 | QE: perturbation in mean anomaly |

```
5700  VK=VK-5.992E-3*UW : QE=QC-(FNM(VK)/PL(IP,4))
```

| 5705 | QD: perturbation in eccentricity |

```
5705  QD=4389*VJ+1129*UV+4262*UU+1089*UW
5710  QD=QD*1E-7
```

| 5715 | QF: perturbation in semi-major axis |

```
5715  QF=8189*UU-817*VJ+781*UW : QF=QF*1E-6
```

| 5720 | Some more arguments ... |

```
5720  VD=SIN(2*JC) : VE=COS(2*JC)
5725  VF=SIN(J8) : VG=COS(J8)
```

| 5730 | QA: perturbation in longitude |

```
5730  QA=-9.556E-3*SIN(JA)-5.178E-3*SIN(JB)
5735  QA=QA+2.572E-3*VD-2.972E-3*VE*VF-2.833E-3*VD*VG
```

| 5740 | QG: perturbation in heliocentric longitude |
| 5745 | QB: perturbation in radius vector |

```
5740  QG=3.36E-4*VE*VF+3.64E-4*VD*VG : QG=FNM(QG)
5745  QB=-40596+4992*COS(JA)+2744*COS(JB)
5750  QB=QB+2044*COS(JC)+1051*VE : QB=QB*1E-6
```

| 5755 | That's all for Neptune |

```
5755  RETURN
```

| 5760 | Perturbations for Pluto ... |

```
5760  REM      Pluto...
```

| 5765 | ... there aren't any because no dynamical theory ... |
| 5770 | ... exists, so issue a warning instead |

```
5765  Q$="Osculating elements: accurate only near epoch"
5770  IF K=1 THEN PRINT : PRINT Q$ : PRINT
5775  RETURN
```

		1	REM
		2	REM Handling program HPLANS
		3	REM
5	Declare arrays PL(8,9) and AP(8)	5	DIM FL(20),ER(20),SW(20),PL(8,9),AP(8)
10	FNI needed by JULDAY	10	DEF FNI(W)=SGN(W)*INT(ABS(W))
15	FNM converts degrees to radians . . .	15	DEF FNM(W)=1.745329252E-2*W
20	. . . and FND converts radians to degrees	20	DEF FND(W)=5.729577951E1*W
25	FNS returns inverse sine	25	DEF FNS(W)=ATN(W/(SQR(1-W*W)+1E-20))
30	. . . and FNC returns inverse cosine	30	DEF FNC(W)=1.570796327-FNS(W)
35	FNU restores near to range $0-2\pi$ (radians)	35	DEF FNU(W)=W-INT(W/6.283185308)*6.283185308
		40	PRINT : PRINT
		45	PRINT "Visual aspects of the planets"
		50	PRINT "----------------------------"
		55	PRINT : PRINT
60	FL(1)=1 if this is not a new date	60	IF FL(1)=1 THEN GOTO 75
65	Get a new date	65	Q$="Calendar date (D,M,Y) .... "
		70	PRINT : PRINT Q$; : INPUT DY,MN,YR
75	Get a planet name . . .	75	Q$="Which planet (full name, upper case) "
		80	PRINT Q$; : INPUT P$: PRINT
85	. . . and call PLANS	85	UT=0 : GOSUB 4500
90	Skip to end if planet name was not recognised . . .	90	IF IP=0 THEN GOTO 305
95	. . . or if date was impossible (JULDAY)	95	IF ER(1)=1 THEN GOTO 305
100	Display results . . .	100	PRINT "Heliocentric coordinates: "; P$
		105	PRINT "------------ ------------" : PRINT
110	. . . in D,M,S form via MINSEC	110	SW(1)=1 : NC=9 : X=FND(L0) : GOSUB 1000
		115	Q$="Ecliptic longitude (D,M,S) .... "
		120	PRINT Q$+OP$
		125	X=FND(S0) : GOSUB 1000
		130	Q$="Ecliptic latitude (D,M,S) .... "
		135	PRINT Q$+OP$
		140	Q$="Radius vector (AU) .... "
145	Display to six decimal places	145	PRINT Q$+" "+STR$(INT(1E6*P0+0.5)/1E6)
		150	PRINT : PRINT "Geocentric coordinates: "; P$
		155	PRINT "---------- ------------" : PRINT
		160	X=FND(EP) : GOSUB 1000
		165	Q$="Ecliptic longitude (D,M,S) .... "
		170	PRINT Q$+OP$
		175	X=FND(BP) : GOSUB 1000
		180	Q$="Ecliptic latitude (D,M,S) .... "
		185	PRINT Q$+OP$
		190	Q$="Distance from Earth (AU) .... "
195	Display to six decimal places	195	PRINT Q$+" "+STR$(INT(1E6*V0+0.5)/1E6)
200	Call NUTAT for nutation and add it to ecliptic long.	200	GOSUB 1800 : EP=EP+FNM(DP)
210	Allow for aberration in longitude . . .	205	A=LG+PI-EP : B=COS(A) : C=SIN(A)
215	. . . and latitude	210	EP=EP-(9.9387E-5*B/COS(BP))
220	Call EQECL for right ascension and declination . . .	215	BP=BP-(9.9387E-5*C*SIN(BP))
		220	X=EP : Y=BP : SW(3)=-1 : GOSUB 2000
		225	P=FND(P) : Q=FND(Q) : X=P/15.0 : GOSUB 1000
		230	Q$="Apparent right ascension (H,M,S) .. "
235	. . . and display in H,M,S form neatly	235	PRINT Q$+" "+MID$(OP$,4,12)
		240	Q$="Apparent declination (D,M,S) ...... "
		245	X=Q : GOSUB 1000 : PRINT Q$+OP$

250	Calculate the solar elongation
260	. . . and display to three decimal places
270	Calculate and display angular diameter to one decimal place
275	The phase . . .
285	. . . displayed with two decimal places
290	The approximate visual magnitude . . .
300	. . . displayed with one decimal place
305	Another go?
310	YESNO gets the answer
320	If FL(1)=0 we have not yet made any calculation
330	Reset all the flags for a new date . . .
335	. . . 1: JULDAY, 6: NUTAT, 7: EQECL

```
250   D=-COS(BP)*COS(EP-LG) : E=FND(FNC(D)) : PRINT
255   Q$="Solar elongation (degrees) ........ "
260   PRINT Q$+"   "+STR$(INT(1E3*E+0.5)/1E3)
265   Q$="Angular size (arcsec) ............ "
270   PRINT Q$+"   "+STR$(INT((10*PL(IP,8)/RH)+0.5)/10)
275   F=0.5*(1.0+COS(EP-PD))
280   Q$="Phase (1=full) ................... "
285   PRINT Q$+"   "+STR$(INT(1E2*F+0.5)/1E2)
290   X=LOG(PV*RH/SQR(F)) : M=5.0*X/LOG(10.0)+PL(IP,9)
295   Q$="Approximate magnitude ............ "
300   PRINT Q$+"   "+STR$(INT(10.0*M+0.5)/10.0)
305   Q$="Again (Y or N) ................... "
310   PRINT : GOSUB 960
315   IF E=0 THEN STOP
320   IF FL(1)=0 THEN GOTO 60
325   Q$="Same date (Y or N) ............... "
330   GOSUB 960 : FL(1)=E : FL(7)=E : FL(6)=E
335   GOTO 60

INCLUDE YESNO, MINSEC, JULDAY, OBLIQ, NUTAT, EQECL,
        ANOMALY, SUN, PELMENT, PLANS
```

4500 PLANS

Example

```
Visual aspects of the planets
-----------------------------


Calendar date (D,M,Y) .............. ? 30,5,1984
Which planet (full name, upper case) ? MERCURY

Heliocentric coordinates: MERCURY
------------ ------------

Ecliptic longitude (D,M,S) ........    - 34 46 20.12
Ecliptic latitude (D,M,S) .........    -  6 57  5.34
Radius vector (AU) ................    0.401741

Geocentric coordinates: MERCURY
---------- ------------

Ecliptic longitude (D,M,S) ........    + 45 55 55.89
Ecliptic latitude (D,M,S) .........    -  2 47 16.47
Distance from Earth (AU) ..........    0.999923
Apparent right ascension (H,M,S) ..       2 57  8.28
Apparent declination (D,M,S) ......    + 13 56  8.81

Solar elongation (degrees) ........    23.015
Angular size (arcsec) .............    6.7
Phase (1=full) ....................    0.58
Approximate magnitude .............    -1.7

Again (Y or N) .................... ? Y
Same date (Y or N) ................ ? Y
Which planet (full name, upper case) ? SATURN

Heliocentric coordinates: SATURN
------------ ------------

Ecliptic longitude (D,M,S) ........    +223 55 33.24
Ecliptic latitude (D,M,S) .........    +  2 19 49.17
Radius vector (AU) ................    9.866205

Geocentric coordinates: SATURN
---------- ------------

Ecliptic longitude (D,M,S) ........    +221 11 41.66
Ecliptic latitude (D,M,S) .........    +  2 34  1.15
Distance from Earth (AU) ..........    8.957211
Apparent right ascension (H,M,S) ..      14 38 19.72
Apparent declination (D,M,S) ......    - 12 44 52.50

Solar elongation (degrees) ........    152.309
Angular size (arcsec) .............    18.5
Phase (1=full) ....................    1
Approximate magnitude .............    0.9

Again (Y or N) .................... ? N
```

The Astronomical Almanac gives the following values:

1984 May 30.0

	Mercury	Saturn
Apparent right ascension	2 57 08.24	14 38 18.136
Apparent declination:	13 56 08.6	−12 44 46.06
True geocentric distance:	0.9999357	8.9567650
Stellar magnitude:	−0.1	0.2
Diameter (arcseconds):	6.73	18.50 (equatorial)
Phase:	0.583	(phase angle = 2.5 degrees)

This handling program displays an ephemeris for any of the eight major planets, that is a list of positions calculated at regular intervals throughout a given period.

Notes		**HPLANEP**

```
           1      REM
           2      REM        Handling program HPLANEP
           3      REM

5  Declare arrays PL(8,9) and AP(8)    5    DIM FL(20),ER(20),SW(20),PL(8,9),AP(8)
10 FNI is needed by JULDAY             10   DEF FNI(W)=SGN(W)*INT(ABS(W))
15 ...and FNL by CALDAY                15   DEF FNL(W)=FNI(W)+FNI((SGN(W)-1.0)/2.0)
20 Degrees to radians ...              20   DEF FNM(W)=1.745329252E-2*W
25 ...and radians to degrees           25   DEF FND(W)=5.729577951E1*W
30 Returns inverse sine ...            30   DEF FNS(W)=ATN(W/(SQR(1-W*W)+1E-20))
35 ...and inverse cosine               35   DEF FNC(W)=1.570796327-FNS(W)
40 FNU reduces near to range 0–2π      40   DEF FNU(W)=W-INT(W/6.283185308)*6.283185308
   (radians)
                                       45   PRINT : PRINT : PRINT "Planet ephemeris"
                                       50   PRINT "----------------" : PRINT
                                       55   PRINT

60 Get the planet's name               60   Q$="Which planet (full name, upper case) "
                                       65   PRINT : PRINT Q$; : INPUT P$

70 FL(1)=0 for new dates               70   IF FL(1)=1 THEN GOTO 105
75 Display the ephemeris from ...       75   Q$="Starting date (D,M,Y) ............. "
                                       80   PRINT : PRINT Q$; : INPUT DA,MA,YA
85 ...every how many days? ...          85   Q$="Step interval (days) .............. "
                                       90   PRINT Q$; : INPUT SD
95 ...for a total of this many times    95   Q$="... for how many steps ............ "
                                      100   PRINT Q$; : INPUT NS : PRINT

105 Display the headline              105   PRINT "Apparent geocentric coordinates at 0h UT"
                                      110   PRINT : PRINT TAB(5);"DATE";TAB(19);"LONGITUDE";
```

4500 PLANS

```
                    115   PRINT TAB(33);"RIGHT ASCENSION";TAB(51);"DECLINATION"
                    120   PRINT TAB(22);"DMS";TAB(38);"HMS";TAB(55);"DMS"
                    125   PRINT "-----------------------------------------------";
                    130   PRINT "--------------------"
                    135   PRINT
```

140 Begin; first convert starting date to Julian date . . .

```
                    140   DY=DA : MN=MA : YR=YA : FL(1)=0
```

145 . . . by calling JULDAY

```
                    145   GOSUB 1100 : DO=DJ
```

150 Impossible date?

```
                    150   IF ER(1)=1 THEN GOTO 250
```

155 Loop for the required number of steps . . .

```
                    155   FOR II=0 TO NS-1 : FL(2)=0
```

160 . . . find calendar date with CALDAY for next DJ . . .

```
                    160   DJ=D0+II*SD : GOSUB 1200 : UT=0 : GOSUB 4500
```

165 . . . call PLANS and abort if planet's name was wrong

```
                    165   IF IP=0 THEN GOTO 250
```

170 Abort if impossible date

```
                    170   IF ER(1)=1 THEN GOTO 250
```

175 . . . call NUTAT for nutation . . .

```
                    175   FL(6)=0 : GOSUB 1800
                    180   SW(1)=1 : NC=9 : AL=FND(EP)
```

185 . . . convert ecliptic longitude to D,M,S (MINSEC) . . .

```
                    185   X=AL : GOSUB 1000 : L$=OP$
```

190 . . . correct ecliptic longitude for nutation . . .

```
                    190   EP=EP+FNM(DP) : A=LG+PI-EP
                    195   B=COS(A) : C=SIN(A)
```

200 . . . and for aberration . . .

```
                    200   EP=EP-(9.9387E-5*B/COS(BP))
```

205 . . . latitude as well . . .

```
                    205   BP=BP-(9.9387E-5*C*SIN(BP)) : X=EP : Y=BP
```

210 . . . find RA and declination with EQECL . . .

```
                    210   SW(3)=-1 : FL(5)=0 : FL(7)=0 : GOSUB 2000
```

215 . . . convert to hours and degrees . . .

```
                    215   P=FND(P) : Q=FND(Q) : X=P/15.0
```

220 . . . H,M,S format via MINSEC

```
                    220   GOSUB 1000 : R$=MID$(OP$,4,12)
                    225   X=Q : GOSUB 1000 : D$=OP$
```

230 . . . and display the results neatly . . .

```
                    230   PRINT DT$;TAB(15);L$;TAB(33);R$;TAB(48);D$
```

235 . . . loop ends

```
                    235   NEXT II
```

240 Finish off the table neatly

```
                    240   PRINT "-----------------------------------------------";
                    245   PRINT "--------------------"
```

250 Another ephemeris?

```
                    250   Q$="Again (Y or N) ....................  "
```

255 Call YESNO to see

```
                    255   PRINT : GOSUB 960
                    260   IF E=0 THEN STOP
```

265 If FL(1)=0 then we have not yet done anything

```
                    265   IF FL(1)=0 THEN GOTO 60
                    270   Q$="Same dates (Y or N) ................  "
```

275 Reset FL(1) for new dates

```
                    275   GOSUB 960 : FL(1)=E : GOTO 60

              INCLUDE YESNO, MINSEC, JULDAY, CALDAY, OBLIQ, NUTAT,
                      EQECL, ANOMALY, SUN, PELMENT, PLANS
```

Example

```
Planet ephemeris
----------------
```

```
Which planet (full name, upper case) ? URANUS

Starting date (D,M,Y) .............. ? 23,10,1992
Step interval (days) ............... ? 10
... for how many steps ............. ? 10

Apparent geocentric coordinates at 0h UT
```

DATE			LONGITUDE DMS			RIGHT ASCENSION HMS			DECLINATION DMS		
23	10	1992	+284	25	28.66	19	2	51.09	− 23	4	30.68
2	11	1992	+284	42	57.12	19	4	6.31	− 23	2	29.10
12	11	1992	+285	4	47.54	19	5	40.34	− 22	59	57.02
22	11	1992	+285	30	31.30	19	7	31.16	− 22	56	55.98
2	12	1992	+285	59	36.38	19	9	36.44	− 22	53	27.67
12	12	1992	+286	31	23.12	19	11	53.26	− 22	49	35.08
22	12	1992	+287	5	11.28	19	14	18.70	− 22	45	21.90
1	1	1993	+287	40	18.14	19	16	49.69	− 22	40	52.42
11	1	1993	+288	15	57.09	19	19	22.90	− 22	36	11.71
21	1	1993	+288	51	24.38	19	21	55.12	− 22	31	25.68

```
Again (Y or N) .................... ? N
```

This handling program calculates the times of rising and setting of any of the major planets anywhere in the world.

Notes		HPLANRS
	1	REM
	2	REM Handling program HPLANRS
	3	REM
5 Declare arrays PL(8,9) and AP(8)	5	DIM FL(20),ER(20),SW(20),PL(8,9),AP(8)
10 FNI is needed by JULDAY ...	10	DEF FNI(W)=SGN(W)*INT(ABS(W))
15 Degrees to radians ...	15	DEF FNM(W)=1.745329252E-2*W -
20 ... and radians to degrees	20	DEF FND(W)=5.729577951E1*W
25 Returns inverse sine ...	25	DEF FNS(W)=ATN(W/(SQR(1-W*W)+1E-20))
30 ... and inverse cosine	30	DEF FNC(W)=1.570796327-FNS(W)
35 FNU reduces near to range 0–2π (radians)	35	DEF FNU(W)=W-INT(W/6.283185308)*6.283185308
	45	PRINT : PRINT
	50	PRINT "Planet rising and setting"
	55	PRINT "------------------------" : PRINT
60 FL(4)=1 indicates same location as before	60	IF FL(4)=1 THEN GOTO 115
65 Get a new location ...	65	Q$="Geog. longitude (D,M,S; W neg.) .... "

4500 PLANS

75	...longitude, GL, in degrees	70	`PRINT Q$; : INPUT XD,XM,XS`
		75	`SW(1)=-1 : GOSUB 1000 : GL=X`
		80	`Q$="Geog. latitude (D,M,S; S neg.) ..... "`
		85	`PRINT Q$; : INPUT XD,XM,XS`
90	...latitude, GP, in radians	90	`SW(1)=-1 : GOSUB 1000 : GP=FNM(X)`
		95	`Q$="Daylight saving (H ahead of zone t)  "`
100	...daylight saving, DS, in hours	100	`PRINT Q$; : INPUT DS`
		105	`Q$="Time zone (hours; West negative) ... "`
110	...time zone, TZ, in hours	110	`PRINT Q$; : INPUT TZ`
115	FL(3)=1 means that the date has not changed	115	`IF FL(3)=1 THEN GOTO 130`
120	Get a new date	120	`Q$="Calendar date (D,M,Y) ............. "`
		125	`PRINT : PRINT Q$; : INPUT DY,MN,YR`
130	Get the planet's name...	130	`Q$="Which planet (full name, upper case) "`
		135	`PRINT : PRINT Q$; : INPUT P$ : PRINT`
140	...and call PLANS	140	`UT=0 : GOSUB 4500`
145	Abort if unrecognised...	145	`IF IP=0 THEN GOTO 375`
150	...or if the date was impossible	150	`IF ER(1)=1 THEN GOTO 375`
155	Call NUTAT and correct for nutation...	155	`GOSUB 1800 : EP=EP+FNM(DP)`
		160	`A=LG+PI-EP : B=COS(A) : C=SIN(A)`
165	...and aberration	165	`EP=EP-(9.9387E-5*B/COS(BP))`
		170	`BP=BP-(9.9387E-5*C*SIN(BP))`
175	Call EQECL for right ascension and declination	175	`X=EP : Y=BP : SW(3)=-1 : GOSUB 2000`
180	Calculate and allow for angular diameter...	180	`A=((PL(IP,8)/2.0)-8.79)/RH : X=P : Y=Q`
185	...and refraction and call RISET	185	`DI=FNM((34.0+A/60.0)/60.0) : GOSUB 3100`
190	Abort if errors in RISET	190	`IF ER(3)<>0 THEN GOTO 375`
195	Display the results...	195	`PRINT : PRINT "Circumstances of rising:"`
200	...for rising...	200	`PRINT "------------------------" : PRINT`
205	...in H,M,S format	205	`X=LU : NC=9 : SW(1)=1 : GOSUB 1000`
		210	`Q$="Local sidereal time (H,M,S) ........ "`
215	...neatly	215	`PRINT Q$+"   "+MID$(OP$,4,12)`
220	Call TIME for the other sorts of time	220	`SW(2)=-1 : TM=LU : GOSUB 1300`
225	Impossible date?	225	`IF ER(1)=1 THEN GOTO 375`
		230	`Q$="Greenwich sidereal time (H,M,S) .... "`
		235	`PRINT Q$+"   "+SG$`
		240	`Q$="Universal time (H,M,S) ............. "`
		245	`PRINT Q$+"   "+UT$`
		250	`Q$="...may be in error by up to 4 minutes"`
255	Announce if conversion was ambiguous (TIME)	255	`IF ER(2)=1 THEN PRINT Q$`
		260	`Q$="Local civil time (H,M,S) ........... "`
		265	`PRINT Q$+"   "+TL$`
270	Azimuths in D,M,S format	270	`X=FND(AU) : SW(1)=1 : GOSUB 1000`
		275	`Q$="Azimuth (D,M,S; zero is North) ..... "`
		280	`PRINT Q$+OP$`
		285	`PRINT : PRINT "Circumstances of setting:"`
290	...for setting...	290	`PRINT "------------------------" : PRINT`
295	...in H,M,S format...	295	`X=LD : NC=9 : SW(1)=1 : GOSUB 1000`
		300	`Q$="Local sidereal time (H,M,S) ........ "`
305	...neatly	305	`PRINT Q$+"   "+MID$(OP$,4,12)`
310	Call TIME for the other sorts of time	310	`SW(2)=-1 : TM=LD : GOSUB 1300`
315	Impossible date?	315	`IF ER(1)=1 THEN GOTO 375`
		320	`Q$="Greenwich sidereal time (H,M,S) .... "`
		325	`PRINT Q$+"   "+SG$`
		330	`Q$="Universal time (H,M,S) ............. "`
		335	`PRINT Q$+"   "+UT$`
		340	`Q$="...may be in error by up to 4 minutes"`
345	Announce if conversion was ambiguous (TIME)	345	`IF ER(2)=1 THEN PRINT Q$`
		350	`Q$="Local civil time (H,M,S) ........... "`

360	Azimuths in D,M,S format	355	`PRINT Q$+" "+TL$`
		360	`X=FND(AD) : SW(1)=1 : GOSUB 1000`
		365	`Q$="Azimuth (D,M,S; zero is North) ..... "`
		370	`PRINT Q$+OP$`
375	Another go?	375	`Q$="Again (Y or N) ..................... "`
380	Call YESNO to see	380	`PRINT : GOSUB 960`
		385	`IF E=0 THEN STOP`
390	If FL(1)=0 we have not yet done	390	`IF FL(1)=0 THEN GOTO 60`
	anything	395	`Q$="Same place (Y or N) ............... "`
400	Set FL(4)=0 for a new place	400	`GOSUB 960 : FL(4)=E`
		405	`Q$="Same date (Y or N) ................ "`
410	Reset the date flags for a new date ...	410	`GOSUB 960 : FL(1)=E : FL(3)=E : FL(6)=E : FL(7)=E`
415	...1: JULDAY, 3: TIME, 6: NUTAT,	415	`PRINT : GOTO 60`
	7: EQECL		

```
INCLUDE YESNO, MINSEC, JULDAY, TIME, OBLIQ, NUTAT, EQECL,
        RISET, ANOMALY, SUN, PELMENT, PLANS
```

Example

```
Planet rising and setting
--------------------------

Geog. longitude (D,M,S; W neg.) .... ? 0,2,10.4
Geog. latitude (D,M,S; S neg.) ..... ? 51,45,0
Daylight saving (H ahead of zone t) ? 0
Time zone (hours; West negative) ... ? 0

Calendar date (D,M,Y) .............. ? 21,12,1993

Which planet (full name, upper case) ? MERCURY


Circumstances of rising:
------------------------

Local sidereal time (H,M,S) ........      13 33 30.13
Greenwich sidereal time (H,M,S) ....      13 33 21.44
Universal time (H,M,S) .............       7 33 49.54
Local civil time (H,M,S) ...........       7 33 49.54
Azimuth (D,M,S; zero is North) .....    +129 45  8.13

Circumstances of setting:
-------------------------

Local sidereal time (H,M,S) ........      21 10 50.83
Greenwich sidereal time (H,M,S) ....      21 10 42.13
Universal time (H,M,S) .............      15  9 55.30
Local civil time (H,M,S) ...........      15  9 55.30
Azimuth (D,M,S; zero is North) .....    +230 14 51.87

Again (Y or N) .................... ? N
```

6000 MOON

This routine calculates the true geocentric ecliptic longitude, latitude and horizontal parallax of the Moon at a given instant.

The Moon's motion in its apparent orbit around the Earth is complex. This is largely because it is moving in the gravitational fields of both the Earth and the Sun, the latter being stronger than the former. When the Moon is on the solar side of the Earth, the gravitational fields of the Sun and the Earth oppose one another, while 14 days later the Moon has moved around to the other side of the Earth and the fields reinforce. Furthermore, the Sun's field is not constant but varies as the Earth moves along its own elliptical orbit about the Sun. It is hardly surprising, therefore, that the Moon does not trace out the simple ellipse expected of one body in closed orbit about another in isolation from everything else.

Such is the complexity of the situation that it is best to represent the Moon's orbit by the sums of series of harmonic terms like

$$p \sin (\omega t) + q \cos (\omega t)$$

just as for planetary perturbations. The amplitudes, p and q, decrease as the angular frequency, ω, increases. Hundreds of terms would be needed to match the accuracy of *The Astronomical Almanac*. Here, we consider the first 53 terms in longitude, 47 in latitude, and 31 in the horizontal parallax to achieve accuracies of about 10, 3, and 0.2 arcseconds respectively. The largest terms can be identified with corrections for specific effects. For example, in the Moon's longitude the term

$$6.28875 \times SIN(MD)$$

arises from the first term of the equation of the centre, the solution to Kepler's equation, while the term

$$1.274018 \times SIN(2\,ME - MD)$$

is called *evection*, and is the principal effect of the variation of the Moon's apparent eccentricity. This analysis rapidly becomes difficult, however, as the physical effects merge together into a mass of ever-decreasing contributions.

In the routine given here, the Moon's mean longitude, ML, mean anomaly, MD, mean elongation, ME, mean distance from its ascending node, MF, the longitude of the ascending node, NA, and the Sun's mean anomaly, MS, are all calculated by means of expressions like

$$A0 + (A1 \times T) + (A2 \times T^2) + (A3 \times T^3).$$

As in other subroutines, accuracy is preserved by keeping the absolute magnitudes of the quantities as small as possible. In particular, the term involving A1 is reduced near to the range 0–360 degrees before being added to the other terms by using the form

$$A0 + B + (A2 \times T^2) + (A3 \times T^3),$$

where

$$B = 360 \times (M1 - INT(M1))$$

$$M1 = A1 \times T.$$

INT can be either the least-integer or truncated-integer function.

Subroutine MOON accepts the instant for which to make the calculations as the date, DY (integer days), MN (months), YR (years), and the Universal Time, UT (hours). Alternatively, DY can include the time expressed as a fraction of 24 hours, and UT is then set to zero. The routine returns the Moon's geocentric ecliptic longitude, MM, latitude, BM, and horizontal parallax, PM. If the time is represented by the fractional part of DY, take care to input it with a sufficient number of decimal places since the Moon moves about half a degree per hour. The apparent coordinates of the Moon at a given place may be found by first correcting MM for nutation (NUTAT, 1800) by adding DP, then converting to equatorial coordinates (EQECL, 2000), and finally allowing for geocentric parallax (PARALLX, 2800). This last correction is particularly important for the Moon as its apparent position may be affected by as much as 1 degree.

The distance of the Moon from the Earth can be found from the horizontal parallax by

$$R = 6378.14/SIN(PM) \text{ kilometres,}$$

and its angular diameter by

$$TH = 3.122512E1 \times SIN(PM) \text{ degrees.}$$

The Moon's phase, K, can be found if the Sun's ecliptic longitude, SR, is known (calculated by SUN, 3400). The equations are:

$$K = (1 + COS(I))/2$$

$$I = 180 - D - ((1.468E{-}1 \times SIN(D) \times (1 - 5.49E{-}2 \times SIN(MD))/(1$$
$$- 1.67E{-}2 \times SIN(MS)))$$

$$COS(D) = COS(MM - SR) \times COS(BM)$$

Routine MOON calls JULDAY (1100) to convert the date to a Julian day number, DJ. It makes no alteration to the value of FL(1) before doing so and the current value of DJ is assumed to be correct if FL(1) = 1.

6000 MOON

Details of MOON
Called by GOSUB 6000.

Calculates MM, the true geocentric ecliptic longitude, BM, the true geocentric ecliptic latitude, and PM, the horizontal parallax of the centre of the Moon, given the instant DY (integer days), MN (months), YR (years), and UT (hours). Alternatively, set UT=0 and add the time as a fraction of a day to DY. All angles in *radians*. Strictly, the instant is one of Ephemeris Time (ET), or Terrestrial Dynamic Time (TDT), but UT is usually a sufficiently-good approximation.

Other routine called: JULDAY (1100).

```
5997 REM
5998 REM      Subroutine MOON
5999 REM
```

6000	Call JULDAY for DJ, and find T (centuries since 1900 January 0.5)
6005	Calculate various trigonometric arguments

```
6000 GOSUB 1100 : T=(DJ/36525.0)+(UT/8.766E5) : T2=T*T

6005 M1=2.732158213E1 : M2=3.652596407E2
6010 M3=2.755455094E1 : M4=2.953058868E1
6015 M5=2.721222039E1 : M6=6.798363307E3
6020 Q=DJ+(UT/24.0) : M1=Q/M1 : M2=Q/M2
6025 M3=Q/M3 : M4=Q/M4 : M5=Q/M5 : M6=Q/M6
6030 M1=360*(M1-INT(M1)) : M2=360*(M2-INT(M2))
6035 M3=360*(M3-INT(M3)) : M4=360*(M4-INT(M4))
6040 M5=360*(M5-INT(M5)) : M6=360*(M6-INT(M6))
```

6045	ML = Moon's mean longitude
6050	MS = Sun's mean anomaly
6055	MD = Moon's mean anomaly
6060	ME = Moon's mean elongation
6065	MF = Moon's mean distance from its ascending node
6070	NA = Longitude of Moon's ascending node
6075	FNM converts degrees to radians

```
6045 ML=2.70434164E2+M1-(1.133E-3-1.9E-6*T)*T2
6050 MS=3.58475833E2+M2-(1.5E-4+3.3E-6*T)*T2
6055 MD=2.96104608E2+M3+(9.192E-3+1.44E-5*T)*T2
6060 ME=3.50737486E2+M4-(1.436E-3-1.9E-6*T)*T2
6065 MF=11.250889+M5-(3.211E-3+3E-7*T)*T2

6070 NA=2.59183275E2-M6+(2.078E-3+2.2E-6*T)*T2

6075 A=FNM(51.2+20.2*T) : S1=SIN(A) : S2=SIN(FNM(NA))
6080 B=346.56+(132.87-9.1731E-3*T)*T
6085 S3=3.964E-3*SIN(FNM(B))
6090 C=FNM(NA+275.05-2.3*T) : S4=SIN(C)
6095 ML=ML+2.33E-4*S1+S3+1.964E-3*S2
6100 MS=MS-1.778E-3*S1
6105 MD=MD+8.17E-4*S1+S3+2.541E-3*S2
6110 MF=MF+S3-2.4691E-2*S2-4.328E-3*S4
6115 ME=ME+2.011E-3*S1+S3+1.964E-3*S2
6120 E=1-(2.495E-3+7.52E-6*T)*T : E2=E*E
6125 ML=FNM(ML) : MS=FNM(MS) : NA=FNM(NA)
6130 ME=FNM(ME) : MF=FNM(MF) : MD=FNM(MD)
```

6135	Calculate Moon's true ecliptic longitude, MM ...

```
6135 L=6.28875*SIN(MD)+1.274018*SIN(2*ME-MD)
6140 L=L+6.58309E-1*SIN(2*ME)+2.13616E-1*SIN(2*MD)
6145 L=L-E*1.85596E-1*SIN(MS)-1.14336E-1*SIN(2*MF)
6150 L=L+5.8793E-2*SIN(2*(ME-MD))
6155 L=L+5.7212E-2*E*SIN(2*ME-MS-MD)+5.332E-2*SIN(2*ME+MD)
6160 L=L+4.5874E-2*E*SIN(2*ME-MS)+4.1024E-2*E*SIN(MD-MS)
6165 L=L-3.4718E-2*SIN(ME)-E*3.0465E-2*SIN(MS+MD)
6170 L=L+1.5326E-2*SIN(2*(ME-MF))-1.2528E-2*SIN(2*MF+MD)
6175 L=L-1.098E-2*SIN(2*MF-MD)+1.0674E-2*SIN(4*ME-MD)
6180 L=L+1.0034E-2*SIN(3*MD)+8.548E-3*SIN(4*ME-2*MD)
6185 L=L-E*7.91E-3*SIN(MS-MD+2*ME)-E*6.783E-3*SIN(2*ME+MS)
6190 L=L+5.162E-3*SIN(MD-ME)+E*5E-3*SIN(MS+ME)
6195 L=L+3.862E-3*SIN(4*ME)+E*4.049E-3*SIN(MD-MS+2*ME)
6200 L=L+3.996E-3*SIN(2*(MD+ME))+3.665E-3*SIN(2*ME-3*MD)
6205 L=L+E*2.695E-3*SIN(2*MD-MS)+2.602E-3*SIN(MD-2*(MF+ME))
6210 L=L+E*2.396E-3*SIN(2*(ME-MD)-MS)-2.349E-3*SIN(MD+ME)
6215 L=L+E2*2.249E-3*SIN(2*(ME-MS))-E*2.125E-3*SIN(2*MD+MS)
6220 L=L-E2*2.079E-3*SIN(2*MS)+E2*2.059E-3*SIN(2*(ME-MS)-MD)
6225 L=L-1.773E-3*SIN(MD+2*(ME-MF))-1.595E-3*SIN(2*(MF+ME))
6230 L=L+E*1.22E-3*SIN(4*ME-MS-MD)-1.11E-3*SIN(2*(MD+MF))
6235 L=L+8.92E-4*SIN(MD-3*ME)-E*8.11E-4*SIN(MS+MD+2*ME)
6240 L=L+E*7.61E-4*SIN(4*ME-MS-2*MD)
6245 L=L+E2*7.04E-4*SIN(MD-2*(MS+ME))
6250 L=L+E*6.93E-4*SIN(MS-2*(MD-ME))
6255 L=L+E*5.98E-4*SIN(2*(ME-MF)-MS)
6260 L=L+5.5E-4*SIN(MD+4*ME)+5.38E-4*SIN(4*MD)
6265 L=L+E*5.21E-4*SIN(4*ME-MS)+4.86E-4*SIN(2*MD-ME)
```

```
                                        6270 L=L+E2*7.17E-4*SIN(MD-2*MS)
                                        6275 MM=ML+FNM(L) : TP=6.283185308
6280  ...make sure that it is the range 0–2π    6280 IF MM<0 THEN MM=MM+TP : GOTO 6280
      radians                           6285 IF MM>TP THEN MM=MM-TP : GOTO 6285

6290  Calculate the Moon's true ecliptic  6290 G=5.128189*SIN(MF)+2.80606E-1*SIN(MD+MF)
      latitude, BM...                   6295 G=G+2.77693E-1*SIN(MD-MF)+1.73238E-1*SIN(2*ME-MF)
                                        6300 G=G+5.5413E-2*SIN(2*ME+MF-MD)+4.6272E-2*SIN(2*ME-MF-MD)
                                        6305 G=G+3.2573E-2*SIN(2*ME+MF)+1.7198E-2*SIN(2*MD+MF)
                                        6310 G=G+9.267E-3*SIN(2*ME+MD-MF)+8.823E-3*SIN(2*MD-MF)
                                        6315 G=G+E*8.247E-3*SIN(2*ME-MS-MF)+4.323E-3*SIN(2*(ME-MD)-MF)
                                        6320 G=G+4.2E-3*SIN(2*ME+MF+MD)+E*3.372E-3*SIN(MF-MS-2*ME)
                                        6325 G=G+E*2.472E-3*SIN(2*ME+MF-MS-MD)
                                        6330 G=G+E*2.222E-3*SIN(2*ME+MF-MS)
                                        6335 G=G+E*2.072E-3*SIN(2*ME-MF-MS-MD)
                                        6340 G=G+E*1.877E-3*SIN(MF-MS+MD)+1.828E-3*SIN(4*ME-MF-MD)
                                        6345 G=G-E*1.803E-3*SIN(MF+MS)-1.75E-3*SIN(3*MF)
                                        6350 G=G+E*1.57E-3*SIN(MD-MS-MF)-1.487E-3*SIN(MF+ME)
                                        6355 G=G-E*1.481E-3*SIN(MF+MS+MD)+E*1.417E-3*SIN(MF-MS-MD)
                                        6360 G=G+E*1.35E-3*SIN(MF-MS)+1.33E-3*SIN(MF-ME)
                                        6365 G=G+1.106E-3*SIN(MF+3*MD)+1.02E-3*SIN(4*ME-MF)
                                        6370 G=G+8.33E-4*SIN(MF+4*ME-MD)+7.81E-4*SIN(MD-3*MF)
                                        6375 G=G+6.7E-4*SIN(MF+4*ME-2*MD)+6.06E-4*SIN(2*ME-3*MF)
                                        6380 G=G+5.97E-4*SIN(2*(ME+MD)-MF)
                                        6385 G=G+E*4.92E-4*SIN(2*ME+MD-MS-MF)+4.5E-4*SIN(2*(MD-ME)-MF)
                                        6390 G=G+4.39E-4*SIN(3*MD-MF)+4.23E-4*SIN(MF+2*(ME+MD))
                                        6395 G=G+4.22E-4*SIN(2*ME-MF-3*MD)-E*3.67E-4*SIN(MS+MF+2*ME-MD)
                                        6400 G=G-E*3.53E-4*SIN(MS+MF+2*ME)+3.31E-4*SIN(MF+4*ME)
                                        6405 G=G+E*3.17E-4*SIN(2*ME+MF-MS+MD)
                                        6410 G=G+E2*3.06E-4*SIN(2*(ME-MS)-MF)-2.83E-4*SIN(MD+3*MF)
                                        6415 W1=4.664E-4*COS(NA) : W2=7.54E-5*COS(C)
                                        6420 BM=FNM(G)*(1.0-W1-W2)

6425  Calculate the Moon's horizontal parallax,  6425 PM=9.50724E-1+5.1818E-2*COS(MD)+9.531E-3*COS(2*ME-MD)
      PM...                             6430 PM=PM+7.843E-3*COS(2*ME)+2.824E-3*COS(2*MD)
                                        6435 PM=PM+8.57E-4*COS(2*ME+MD)+E*5.33E-4*COS(2*ME-MS)
                                        6440 PM=PM+E*4.01E-4*COS(2*ME-MD-MS)
                                        6445 PM=PM+E*3.2E-4*COS(MD-MS)-2.71E-4*COS(ME)
                                        6450 PM=PM-E*2.64E-4*COS(MS+MD)-1.98E-4*COS(2*MF-MD)
                                        6455 PM=PM+1.73E-4*COS(3*MD)+1.67E-4*COS(4*ME-MD)
                                        6460 PM=PM-E*1.11E-4*COS(MS)+1.03E-4*COS(4*ME-2*MD)
                                        6465 PM=PM-8.4E-5*COS(2*MD-2*ME)-E*8.3E-5*COS(2*ME+MS)
                                        6470 PM=PM+7.9E-5*COS(2*ME+2*MD)+7.2E-5*COS(4*ME)
                                        6475 PM=PM+E*6.4E-5*COS(2*ME-MS+MD)-E*6.3E-5*COS(2*ME+MS-MD)
                                        6480 PM=PM+E*4.1E-5*COS(MS+ME)+E*3.5E-5*COS(2*MD-MS)
                                        6485 PM=PM-3.3E-5*COS(3*MD-2*ME)-3E-5*COS(MD+ME)
                                        6490 PM=PM-2.9E-5*COS(2*(MF-ME))-E*2.9E-5*COS(2*MD+MS)
                                        6495 PM=PM+E2*2.6E-5*COS(2*(ME-MS))-2.3E-5*COS(2*(MF-ME)+MD)
                                        6500 PM=PM+E*1.9E-5*COS(4*ME-MS-MD)
6505  Return from MOON                  6505 PM=FNM(PM) : RETURN
```

	1 REM
	2 REM Handling program HMOON
	3 REM
10 FNI is truncated-integer function needed by JULDAY	5 DIM FL(20),ER(20),SW(20) 10 DEF FNI(W)=SGN(W)*INT(ABS(W))
15 FNM – degrees to radians conversion . . .	15 DEF FNM(W)=1.745329252E-2*W
20 . . . and FND – radians to degrees	20 DEF FND(W)=5.729577951E1*W
25 FNS returns the inverse sine	25 DEF FNS(W)=ATN(W/(SQR(1-W*W)+1E-20))
	30 PRINT : PRINT : SP=0 : SH=0
	35 PRINT "The position of the Moon"
	40 PRINT "------------------------" : PRINT
	45 PRINT : PRINT
50 SP is a local flag meaning 'same place' if set to 1	50 IF SP=1 THEN GOTO 75
55 Get details of new place . . .	55 Q$="Daylight saving (H ahead of zone t) "
60 . . . DS is daylight saving correction (hours)	60 PRINT Q$; : INPUT DS
	65 Q$="Time zone (hours; West negative) ... "
70 . . . TZ is time zone in hours	70 PRINT Q$; : INPUT TZ
75 FL(1)=1 means that the date has not changed	75 IF FL(1)=1 THEN GOTO 90
	80 Q$="Calendar date (D,M,Y) .... "
85 Get a new date	85 PRINT : PRINT Q$; : INPUT DY,MN,YR
90 SH is a local flag meaning 'same time' when set to 1	90 IF SH=1 THEN GOTO 110
	95 Q$="Local civil time (H,M,S) .... "
100 Get a new local civil time . . .	100 PRINT : PRINT Q$; : INPUT XD,XM,XS
105 . . . in hours via MINSEC	105 SW(1)=-1 : GOSUB 1000 : TM=X
110 Set value of UT and call MOON, then NUTAT	110 UT=TM-DS-TZ : GOSUB 6000 : GOSUB 1800
115 Was the date impossible?	115 IF ER(1)=1 THEN GOTO 255
120 Display the results . . .	120 PRINT : PRINT "Apparent geocentric coordinates:"
	125 PRINT "-------------------------------" : PRINT
130 Correct lunar longitude for nutation and convert . . .	130 SW(1)=1 : AL=FND(MM)+DP : X=AL : GOSUB 1000
135 . . . to D,M,S format with MINSEC	135 Q$="Ecliptic longitude (D,M,S) .... "
140 Display the longitude . . .	140 PRINT Q$+OP$: X=FND(BM) : GOSUB 1000 : SW(3)=-1
	145 Q$="Ecliptic latitude (D,M,S) .... "
150 . . . and latitude; call EQECL to get right ascension and declination . . .	150 PRINT Q$+OP$: X=FNM(AL) : Y=BM : GOSUB 2000
155 . . . convert right ascension to H,M,S . . .	155 P=FND(P) : Q=FND(Q) : X=P/15.0 : GOSUB 1000
	160 Q$="Right ascension (H,M,S) .... "
165 . . . and display it neatly	165 PRINT Q$+" "+MID$(OP$,4,12) : X=Q : GOSUB 1000
	170 Q$="Declination (D,M,S) .... "
175 Display the declination in D,M,S format	175 PRINT Q$+OP$: PRINT
180 Convert parallax to D,M,S format with MINSEC	180 X=FND(PM) : GOSUB 1000
	185 Q$="Horizontal parallax (D,M,S) .... "
190 Find the Earth–Moon distance from the parallax	190 PRINT Q$+OP$: R=6378.14/SIN(PM)
195 Find the angular diameter TH	195 F=R/384401.0 : TH=5.181E-1/F
	200 Q$="Earth-Moon distance (km) .... "
	205 PRINT Q$+" "+STR$(INT(R+0.5)) : X=TH : GOSUB 1000
	210 Q$="Angular diameter (D,M,S) .... "
215 Call SUN, and then calculate the phase, F . . .	215 PRINT Q$+OP$: GOSUB 3400 : CD=COS(MM-SR)*COS(BM)
	220 D=1.570796327-FNS(CD) : SD=SIN(D)
	225 I=1.468E-1*SD*(1.0-5.49E-2*SIN(MD))
	230 I=I/(1.0-1.67E-2*SIN(MS)) ☛

6000 MOON

```
235   I=PI-D-FNM(I) : K=(1.0+COS(I))/2
240   FM=INT(K*1000+0.5)/1000
245   Q$="Phase of Moon (full = 1) .......... "
250   PRINT Q$+"   "+STR$(FM)

255   Q$="Again (Y or N) .................... "
260   PRINT : GOSUB 960
265   IF E=0 THEN STOP
270   Q$="Same place (Y or N) ............... "
275   GOSUB 960 : SP=E
280   Q$="Same date (Y or N) ................ "
285   GOSUB 960 : FL(1)=E : FL(7)=E : FL(6)=E
290   Q$="Same time (Y or N) ................ "
295   GOSUB 960 : SH=E : PRINT : GOTO 50

INCLUDE YESNO, MINSEC, JULDAY, OBLIQ, NUTAT, EQECL,
        ANOMALY, SUN, MOON
```

The Astronomical Almanac gives the following values:

Moon	25th February 1984
Apparent right ascension:	17 19 21.0
Apparent declination:	−23 52 23
Horizontal parallax:	00 56 03.1
Semi diameter:	00 15 16.39
Fraction illuminated:	0.37

Example

```
The position of the Moon
------------------------


Daylight saving (H ahead of zone t)   ? 0
Time zone (hours; West negative) ...  ? 0

Calendar date (D,M,Y) .............  ? 25,2,1984

Local civil time (H,M,S) ..........  ? 0,0,0

Apparent geocentric coordinates:
-------------------------------

Ecliptic longitude (D,M,S) ........    +260 42 46.34
Ecliptic latitude (D,M,S) .........    -   0 45 28.16
Right ascension (H,M,S) ...........      17 19 20.65
Declination (D,M,S) ...............    - 23 52 22.51

Horizontal parallax (D,M,S) .......    +  0 56  3.15
Earth-Moon distance (km) ..........    391194
Angular diameter (D,M,S) ..........    +  0 30 32.77
Phase of Moon (full = 1) ..........    0.371

Again (Y or N) ....................  ? Y
Same place (Y or N) ...............  ? N
Same date (Y or N) ................  ? N
Same time (Y or N) ................  ? N

Daylight saving (H ahead of zone t)  ? 1
Time zone (hours; West negative) ...  ? 6

Calendar date (D,M,Y) .............  ? 14,8,1989

Local civil time (H,M,S) ..........  ? 11,42,20

Apparent geocentric coordinates:
-------------------------------

Ecliptic longitude (D,M,S) ........    +283 24  1.33
Ecliptic latitude (D,M,S) .........    -  3 34  4.63
Right ascension (H,M,S) ...........      18 59 49.08
Declination (D,M,S) ...............    - 26 19  3.37

Horizontal parallax (D,M,S) .......    +  0 57 29.38
Earth-Moon distance (km) ..........    381416
Angular diameter (D,M,S) ..........    +  0 31 19.76
Phase of Moon (full = 1) ..........    0.894

Again (Y or N) ....................  ? N
```

```
1    REM
2    REM          Handling program HMOONEP
3    REM

5    DIM FL(20),ER(20),SW(20)
10   DEF FNI(W)=SGN(W)*INT(ABS(W))

15   DEF FNL(W)=FNI(W)+FNI((SGN(W)-1.0)/2.0)

20   DEF FNM(W)=1.745329252E-2*W
25   DEF FND(W)=5.729577951E1*W
30   DEF FNS(W)=ATN(W/(SQR(1-W*W)+1E-20))
35   PRINT : PRINT : PRINT "Lunar ephemeris"
40   PRINT "---------------" : PRINT
45   PRINT

50   IF SP=1 THEN GOTO 75

55   Q$="Daylight saving (H ahead of zone t)    "
60   PRINT Q$; : INPUT DS
65   Q$="Time zone (hours; West negative) ... "
70   PRINT Q$; : INPUT TZ

75   Q$="Starting local date (D,M,Y) ........ "
80   PRINT : PRINT Q$; : INPUT DA,MA,YA
85   Q$="Starting local civil time (H,M,S) .. "
90   PRINT Q$; : INPUT XD,XM,XS
95   SW(1)=-1 : GOSUB 1000 : TM=X
100  Q$="Step interval (hours) ............. "
105  PRINT Q$; : INPUT SD
110  Q$="... for how many steps ............ "
115  PRINT Q$; : INPUT NS

120  PRINT : PRINT "Apparent geocentric coordinates:"
125  PRINT "--------------------------------" : PRINT
130  PRINT TAB(7);"Local";TAB(17);"Local civil";
135  PRINT TAB(33);"Right ascension";TAB(51);"Declination"
140  PRINT TAB(7);"date";TAB(20);"time";TAB(38);"HMS";
145  PRINT TAB(55);"DMS" : PRINT "--------------------";
150  PRINT "---------------------------------------"
155  PRINT

160  DY=DA : MN=MA : YR=YA : FL(1)=0
165  GOSUB 1100 : D0=DJ : U0=TM-DS-TZ

170  IF ER(1)=1 THEN GOTO 250
175  FOR II=0 TO NS-1 : FL(2)=0
180  DJ=D0+(U0+II*SD)/24.0 : GOSUB 1200 : UT=0

185  GOSUB 6000 : FL(6)=0 : GOSUB 1800

190  SW(1)=1 : NC=9 : X=MM+FNM(DP) : Y=BM
195  SW(3)=-1 : FL(5)=0 : FL(7)=0 : GOSUB 2000

200  P=FND(P) : Q=FND(Q) : X=P/15.0
205  GOSUB 1000 : R$=MID$(OP$,4,12) : FL(2)=0
210  X=Q : GOSUB 1000 : DE$=OP$ : TM=FD*24.0+DS+TZ
215  IF TM>24.0 THEN TM=TM-24.0 : DJ=DJ+1 : GOSUB 1200
220  IF TM<0 THEN TM=TM+24.0 : DJ=DJ-1 : GOSUB 1200
225  X=TM : GOSUB 1000 : L$=MID$(OP$,4,12)
```

Notes (left column):

10 FNI is truncated-integer function needed by JULDAY
15 FNL is least-integer function needed by CALDAY
20 FNM – degrees to radians conversion . . .
25 . . . and FND – radians to degrees
30 FNS returns the inverse sine

50 SP is a local switch meaning 'same place' when set to 1
55 Get details of a new place . . .
60 . . . DS is daylight saving correction in hours
70 . . . TZ is time zone in hours

75 Get the date at which to begin the ephemeris . . .
85 . . . and the local civil time . . .

95 . . . in hours via MINSEC; result in TM

105 Step interval SD in hours . . .

115 . . . and the length of the list

120 Display the banner headlines

160 Here we go . . .
165 . . . find DJ of start via JULDAY, and set initial UT . . .
170 Allow for numskulls!
175 . . . set up the loop (index II) . . .
180 . . . find next value of DJ and convert to calendar . . .
185 . . . date with CALDAY; call MOON; call NUTAT
190 . . . allow for nutation (DP) . . .
195 . . . and call EQECL for right ascension and declination . . .
200 . . . convert results to minutes and seconds format . . .
210 . . . find the next local civil time . . .
215 . . . allowing for change of date if necessary . . .

6000 MOON

230	. . . and display the results . . .	230	`PRINT DT$;TAB(16);L$;TAB(33);R$;TAB(48);DE$`
235	. . . end of the loop	235	`NEXT II`
240	Finish off list neatly	240	`PRINT "---------------------------------------------";`
		245	`PRINT "---------------------"`
250	Another ephemeris?	250	`Q$="Again (Y or N) ....................  "`
255	YESNO has the answer	255	`PRINT : GOSUB 960`
		260	`IF E=0 THEN STOP`
		265	`Q$="Same place (Y or N) ...............  "`
270	SP=0 for a new place	270	`GOSUB 960 : SP=E : PRINT : GOTO 50`

```
INCLUDE YESNO, MINSEC, JULDAY, CALDAY, OBLIQ, NUTAT,
        EQECL, ANOMALY, SUN, MOON
```

This handling program displays an ephemeris of the Moon, i.e. a list of its positions calculated at regular intervals throughout a given period.

6000 MOON

Example

```
Lunar ephemeris
---------------


Daylight saving (H ahead of zone t)  ? 1
Time zone (hours; West negative) ... ? -5

Starting local date (D,M,Y) ........ ? 19,4,1991
Starting local civil time (H,M,S) .. ? 22,30,0
Step interval (hours) .............. ? 0.5
... for how many steps ............. ? 10

Apparent geocentric coordinates:
--------------------------------
```

Local date	Local civil time	Right ascension HMS	Declination DMS
19 4 1991	22 30 0.00	6 47 34.68	+ 24 2 7.80
19 4 1991	23 0 0.00	6 48 51.79	+ 23 59 7.15
19 4 1991	23 30 0.00	6 50 8.81	+ 23 56 4.15
20 4 1991	0 0 0.00	6 51 25.74	+ 23 52 58.80
20 4 1991	0 30 0.00	6 52 42.60	+ 23 49 51.11
20 4 1991	1 0 0.00	6 53 59.37	+ 23 46 41.09
20 4 1991	1 30 0.00	6 55 16.06	+ 23 43 28.76
20 4 1991	2 0 0.00	6 56 32.66	+ 23 40 14.12
20 4 1991	2 30 0.00	6 57 49.18	+ 23 36 57.18
20 4 1991	3 0 0.00	6 59 5.61	+ 23 33 37.95

```
Again (Y or N) .................... ? N
```

6600 MOONRS

This routine calculates the universal times, local civil times and azimuths of moonrise and moonset anywhere on the Earth's surface for any date

Routine RISET (3100) calculates the times and azimuths of rising and setting of a celestial body given its right ascension and declination as input parameters. We may therefore combine it with the routines MOON (6000) and EQECL (2000) to find the circumstances of moonrise and moonset at any place on the Earth's surface for any given calendar date. There is a snag, however, as the Moon's position is constantly changing by about half a degree of ecliptic longitude per hour and, unless we already know the times of rising and setting, we cannot find the correct position of the Moon in advance. If we use the position calculated for a set time, say midnight, we may be as much as one hour out in the calculation of moonrise and moonset. We must therefore adopt an iterative procedure, refining an initial crude estimate to achieve the required accuracy, just as in SUNRS (3600).

Routine MOONRS first calculates the right ascension and declination of the Moon at local midday of the given date, making allowance for nutation (NUTAT, 1800). Then the universal times (UTs) and azimuths of the corresponding moonrise and moonset are found via RISET (3100) and TIME (1300) with the corrections for parallax, refraction and the angular size of the Moon's disk incorporated in the parameter DI. The Moon's position is now recalculated at each of these approximations to moonrise or moonset and the whole procedure repeated twice more to find times of rising and setting accurate to within one minute, and azimuths to within about 1 arcminute. Lines 6700–6720 form the core of the routine, returning the local sidereal times of rising and setting for a given value of DJ, the number of Julian days since 1900 January 0.5. The calculations are for the upper limb of the Moon and they assume the standard horizontal refraction of 34 arcminutes.

MOONRS returns the UTs of moonrise in the string variable UU$, and of moonset in UD$, the local civil times, TU$ and TD$, and the corresponding aximuths, AU and AD (radians), on the date given by DY, MN, YR, and at the place specified by the geographical longitude GL (degrees), latitude GP (radians), time zone TZ (hours) and daylight saving correction DS (hours).

6600 MOONRS

Since the Moon's position changes so rapidly throughout the day, it is sometimes possible that either one of the events moonrise or moonset does not occur on the given date, but happens early the next day. Error flags ER(6) and ER(7) indicate when this happens for moonrise and moonset respectively. Normally they are set to 0 by the routine (line 6610), but are set to 1 if there is no event. The error flag ER(3) or RISET (3100) indicates whether the Moon never rises (ER(3) = 1) or never sets (ER(3) = −1) on the given date.

Routine MOONRS has a great deal of calculating to do, and is therefore quite slow to execute. On my machine it takes about half a minute to return from a call with the results; yours may be faster or slower depending on the precision of your BASIC and the clock rate of your microprocessor. There are several PRINT statements in the routine which display '*' or a number from time to time as the calculations progress. If you find these a distraction, remove *all* the PRINT statements (i.e. in lines 6605, 6615, 6635, 6645, 6675, 6720).

Details of MOONRS

Called by GOSUB 6600.

Calculates the UTs of moonrise and moonset, returning the values in string format in the variables UU$ and UD$ respectively, the local civil times TU$ and TD$, and the corresponding azimuths AU and AD (radians). Input parameters are the date, DY (*integer* days), MN (months), YR (years), the geographical longitude, GL (*degrees*), latitude, GP (*radians*), time zone, TZ (hours ahead of UT), and daylight saving correction, DS (hours added to zone time).

ER(6) is set to 1 by the routine if moonrise does not occur until the next day. ER(7) is set to 1 if moonset does not occur until the next day.

Both ER(6) and ER(7) are set to 0 otherwise.

Other routines called: TIME (1300), NUTAT (1800), EQECL (2000), RISET (3100), MOON (6000).

		6597 REM
		6598 REM Subroutine MOONRS
		6599 REM
6600	Set UT to local midday, and call local riset routine	6600 SW(2)=-1 : UT=12.0+TZ+DS : GOSUB 6700
6605	Impossible date, or Moon never crosses horizon?	6605 IF ER(1)=1 OR ER(3)<>0 THEN PRINT : RETURN
6610	LU, LD are local sidereal times of rising and setting	6610 ER(6)=0 : ER(7)=0 : LA=LU : LB=LD
6615	Set up iterative loop; display progress through it	6615 FOR K=1 TO 3 : PRINT K;
6620	Convert to UT using TIME for rising . . .	6620 G1=GU : TM=LA : GOSUB 1300 : GU=UT
6625	. . . and setting	6625 G2=GD : TM=LB : GOSUB 1300 : GD=UT
6630	Now use latest UT to find better time of rising . . .	6630 UT=GU : GOSUB 6680 : LA=LU : AA=AU
6635	Does Moon cross the horizon?	6635 IF ER(3)<>0 THEN PRINT : RETURN
6640	. . . and setting	6640 UT=GD : GOSUB 6680 : LB=LD : AB=AD
6645	Does Moon cross the horizon?	6645 IF ER(3)<>0 THEN PRINT : RETURN
6650	End of iterative loop; return latest azimuths	6650 NEXT K : AU=AA : AD=AB
6655	Did moonrise occur the next day?	6655 IF ABS(GU-G1)>6.0 THEN ER(6)=1
6660	Call TIME to convert to UT and local civil time	6660 TM=LA : GOSUB 1300 : UU$=UT$: TU$=TL$
6665	Did moonset occur the next day?	6665 IF ABS(GD-G2)>6.0 THEN ER(7)=1
6670	Call TIME to convert to UT and local civil time	6670 TM=LB : GOSUB 1300 : UD$=UT$: TD$=TL$
6675	Normal return from MOONRS	6675 PRINT : RETURN
6680	Allow for the local date . . .	6680 DN=DJ : A=UT+TZ+DS
6685	. . . being different from the Greenwich date	6685 IF A>24 THEN DJ=DJ-1
		6690 IF A<0 THEN DJ=DJ+1
		6695 GOSUB 6700 : DJ=DN : RETURN
6700	Local riset routine; call MOON and NUTAT	6700 GOSUB 6000 : GOSUB 1800
6705	Impossible date?	6705 IF ER(1)=1 THEN RETURN
6710	Set DI allowing for angular radius, refraction, and parallax	6710 TH=2.7249E-1*SIN(PM) : DI=TH+9.8902E-3-PM
6715	Add in nutation, and call EQECL for RA and declination	6715 X=MM+FNM(DP) : Y=BM : SW(3)=-1 : GOSUB 2000
6720	Call RISET for local sidereal times, and return	6720 X=P : Y=Q : GOSUB 3100 : PRINT "*"; : RETURN

	1	REM
	2	REM Handling program HMOONRS
	3	REM
	5	DIM FL(20),ER(20),SW(20)
10 FNI is needed by JULDAY	10	DEF FNI(W)=SGN(W)*INT(ABS(W))
15 Converts degrees to radians . . .	15	DEF FNM(W)=1.745329252E-2*W
20 . . . and radians to degrees	20	DEF FND(W)=5.729577951E1*W
25 Returns the inverse sine . . .	25	DEF FNS(W)=ATN(W/(SQR(1-W*W)+1E-20))
30 . . . and inverse cosine	30	DEF FNC(W)=1.570796327-FNS(W)
	35	PRINT : PRINT
	40	PRINT "Moonrise and moonset"
	45	PRINT "---------------------" : PRINT
50 If FL(4)=1 the location is the same as before	50	IF FL(4)=1 THEN GOTO 105
55 Get details of new location . . .	55	Q$="Geog. longitude (D,M,S; W neg.) .... "
	60	PRINT Q$; : INPUT XD,XM,XS
65 . . . longitude GL in degrees via MINSEC	65	SW(1)=-1 : GOSUB 1000 : GL=X
	70	Q$="Geog. latitude (D,M,S; S neg.) ..... "
	75	PRINT Q$; : INPUT XD,XM,XS
80 . . . latitude GP in radians	80	SW(1)=-1 : GOSUB 1000 : GP=FNM(X)
	85	Q$="Daylight saving (H ahead of zone t) "
90 . . . daylight saving correction DS in hours	90	PRINT Q$; : INPUT DS
	95	Q$="Time zone (hours; West negative) ... "
100 . . . and the time zone TZ in hours	100	PRINT Q$; : INPUT TZ
105 If FL(3)=1 the date is the same as before	105	IF FL(3)=1 THEN GOTO 120
110 Get a new date (remember integer days)	110	Q$="Calendar date (D,M,Y) .... "
	115	PRINT : PRINT Q$; : INPUT DY,MN,YR
120 Call MOONRS, and wait . . .	120	PRINT : GOSUB 6600
125 Was the date impossible?	125	IF ER(1)=1 OR ER(3)<>0 THEN GOTO 240
130 Display the results for moonrise . . .	130	PRINT : PRINT "Circumstances of moonrise:"
	135	PRINT "-------------------------"
	140	Q$="++ moonrise occurs next day"
145 . . . if there were any	145	IF ER(6)=1 THEN PRINT : PRINT Q$: GOTO 185
	150	Q$="Universal time (H,M,S) .... "
	155	PRINT : PRINT Q$+UU$
	160	Q$="Local civil time (H,M,S) .... "
	165	PRINT Q$+TU$
170 Azimuth in D,M,S format via MINSEC	170	X=FND(AU) : SW(1)=1 : GOSUB 1000
	175	Q$="Azimuth (D,M,S; zero is North) ..... "
	180	PRINT Q$+OP$
185 Display the results for moonset . . .	185	PRINT : PRINT "Circumstances of moonset:"
	190	PRINT "-------------------------"
	195	Q$="++ moonset occurs next day"
200 . . . if there were any	200	IF ER(7)=1 THEN PRINT : PRINT Q$: GOTO 240
	205	Q$="Universal time (H,M,S) .... "
	210	PRINT : PRINT Q$+UD$
	215	Q$="Local civil time (H,M,S) .... "
	220	PRINT Q$+TD$
225 Azimuth in D,M,S format via MINSEC	225	X=FND(AD) : SW(1)=1 : GOSUB 1000
	230	Q$="Azimuth (D,M,S; zero is North) ..... "
	235	PRINT Q$+OP$
240 Another go?	240	Q$="Again (Y or N) .... "
245 Call YESNO to find out	245	PRINT : GOSUB 960
	250	IF E=0 THEN STOP

168

255	If FL(3)=0 we did not make any calculations	255	IF FL(3)=0 THEN FL(4)=0 : PRINT : GOTO 50

255 If FL(3)=0 we did not make any calculations

265 Set FL(4)=0 for a new place

275 Reset flags for new date: 1: JULDAY, 3: TIME, ...

280 ...6: NUTAT, and 7: EQECL

```
255   IF FL(3)=0 THEN FL(4)=0 : PRINT : GOTO 50
260   Q$="Same place (Y or N) ................ "
265   GOSUB 960 : FL(4)=E
270   Q$="Same date (Y or N) ................. "
275   GOSUB 960 : FL(1)=E : FL(3)=E : FL(6)=E

280   FL(7)=E : PRINT : GOTO 50

INCLUDE YESNO, MINSEC, JULDAY, TIME, OBLIQ, NUTAT,
        EQECL, RISET, MOON, MOONRS
```

6600 MOONRS

Example

```
Moonrise and moonset
--------------------

Geog. longitude (D,M,S; W neg.) .... ? 0,0,0
Geog. latitude (D,M,S; S neg.) ..... ? 30,0,0
Daylight saving (H ahead of zone t)  ? 0
Time zone (hours; West negative) ... ? 0

Calendar date (D,M,Y) .............. ? 7,1,1984

* 1 ** 2 ** 3 **

Circumstances of moonrise:
--------------------------

Universal time (H,M,S) .............    9 58  1.46
Local civil time (H,M,S) ...........    9 58  1.46
Azimuth (D,M,S; zero is North) .....   +107 44  2.83

Circumstances of moonset:
-------------------------

Universal time (H,M,S) .............   21  8 59.44
Local civil time (H,M,S) ...........   21  8 59.44
Azimuth (D,M,S; zero is North) .....   +254 34  8.13

Again (Y or N) .................... ? Y
Same place (Y or N) ............... ? Y
Same date (Y or N) ................ ? N


Calendar date (D,M,Y) .............. ? 11,1,1984

* 1 ** 2 ** 3 **

Circumstances of moonrise:
--------------------------

Universal time (H,M,S) .............   11 54 54.02
Local civil time (H,M,S) ...........   11 54 54.02
Azimuth (D,M,S; zero is North) .....   + 84 54 48.14

Circumstances of moonset:
-------------------------

++ moonset occurs next day

Again (Y or N) .................... ? Y
Same place (Y or N) ............... ? N
Same date (Y or N) ................ ? N

Geog. longitude (D,M,S; W neg.) .... ? -77,0,0
Geog. latitude (D,M,S; S neg.) ..... ? 38,55,0
Daylight saving (H ahead of zone t)  ? 0
Time zone (hours; West negative) ... ? -5

Calendar date (D,M,Y) .............. ? 1,2,1984

* 1 ** 2 ** 3 **
```

```
Circumstances of moonrise:
--------------------------

Universal time (H,M,S) .............     12 23 25.21
Local civil time (H,M,S) ...........      7 23 25.21
Azimuth (D,M,S; zero is North) .....    +119 32 44.23

Circumstances of moonset:
-------------------------

Universal time (H,M,S) .............     22  9 13.51
Local civil time (H,M,S) ...........     17  9 13.51
Azimuth (D,M,S; zero is North) .....    +241 55  7.93

Again (Y or N) .....................   ? N
```

The Astronomical Almanac gives the following values:

Date	Longitude	Latitude	Moonrise	Moonset
7th January 1984	00.00	30.00	09 58	21 09
11th January 1984	00.00	30.00	11 55	(Moon sets on 12th)
1st February 1984	−77.00 (Washington DC)	38.92	07 23	17 09

6800 MOONNF

This routine calculates the instants of new moon and full moon nearest to a given calendar date.

It is sometimes necessary to calculate the date and time of new or full moon corresponding to a particular calendar month. For example, since a solar eclipse can only occur at new moon, and a lunar eclipse at full moon, we need those times to predict the occurrence of an eclipse. This information could be found using the routine MOON (6000) and its handling program, but it would be a tedious exercise as the phase would have to be calculated several times in an iterative procedure of successive approximation. Routine MOONNF adopts the algorithm given by Meeus in his excellent book *Astronomical Formulae for Calculators* (second edition) to calculate the instants of new and full moon to within an accuracy of about 2 minutes.

The date is input as usual via the variables DY, MN, and YR. The instants of new and full moon nearest to the date are calculated in terms of the number of Julian days since 1900 January 0.5, being returned by the pairs of variables FI, FF for full moon and NI, NF for new moon. The first variable in each pair indicates the integer number of Julian days of the event since 1900 January 0.5, while the second variable indicates the fraction of the day. The Julian date of new moon is therefore

JD = 2 415 020 + NI + NF

since 1900 January 0.5 corresponds to Julian date 2 415 020.0. Note that Julian days begin at noon, a fact which must be considered when converting to calendar date and time.

MOONNF makes use of the least-integer function to find the nearest integer which is equal to or less than a given decimal number. Many forms of BASIC represent this function by INT, but some use INT for the truncated-integer function. MOONNF uses its own least-integer function, FNL, which must be defined before calling the routine (line 15 of the handling program). FNL is also used by CALDAY (1200).

Routine MOONNF also returns the value of the Moon's argument of latitude in NB (new moon) and FB (full moon). This quantity is useful in determining whether or not an eclipse is likely to occur and is used by ECLIPSE (7000).

Formulae

$\quad$ NI or FI $= E1 + B1$

$\quad$ NF or FF $= B - B1$

$\quad$ E1 $= INT(E)$

$\quad$ B1 $= INT(B)$

$\quad$ E $= 29.53\ K$

$\quad$ B $= DD + E - E1 + (5.8868E–4 \times K) + (1.178E–4 \times T^2) - (1.55E–7 \times T^3)$
$\qquad + (3.3–4 \times SIN(C)) + 7.5933E–1$

$\quad$ C $= 166.56 + (132.87 \times T) - (9.173E–3 \times T^2)$ degrees

$\quad$ DD $=$ sum of series of terms (see BASIC listing) using the following:

$\quad$ A1 $= 359.2242 + (29.10535608 \times K) - (3.33E–5 \times T^2)$
$\qquad - (3.47E–6 \times T^3)$ degrees

$\quad$ A2 $= 306.0253 + (385.81691806 \times K) + (1.07306E–2 \times T^2)$
$\qquad + (1.236E–5 \times T^3)$ degrees

$\quad$ F $= 21.2964 + (390.67050646 \times K) - (1.6528E–3 \times T^2)$
$\qquad - (2.39E–6 \times T^3)$ degrees

$\quad$ K1 $= (Y0 - 1900 + ((DJ - J0)/365)) \times 12.3685$

$\quad$ K $= FNL(K1 + 0.5)$ for new moon

$\quad$ K $= FNL(K1 + 0.5) + 0.5$ for full moon

$\quad$ T $= K/1236.85$

$\quad$ J0 $=$ Julian days since 1900 January 0.5 to beginning of year

$\quad$ DJ $=$ Julian days since 1900 January 0.5 to date.

Details of MOONNF

Called by GOSUB 6800.

Calculates the instant of new Moon, NI and NF (days), and the instant of full moon, FI and FF (days), nearest to the given calendar date DY (days), MN (months), YR (years). NI and FI are the integer numbers of Julian days since 1900 January 0.5; NF and FF are the fractions of a day. The argument of the Moon's latitude is also returned by NB (new moon, radians) and FB (full moon, radians).

Other routine called: JULDAY (1100).

```
6797  REM
6798  REM        Subroutine MOONNF
6799  REM
```

6800 Save the input date ...
6805 ... making an allowance for BC years
6810 Call JULDAY for DJ of beginning of year
6815 Impossible date?
6820 Call JULDAY again for DJ of date
6825 This should never occur
6830 Now calculate the arguments for local routine ...
6840 ... calling it for new moon ...
6845 and for full moon
6850 Normal return from MOONNF

```
6800  DO=DY : MO=MN : YO=YR
6805  IF YO<0 THEN YO=YO+1
6810  MN=1 : DY=0 : FL(1)=0 : GOSUB 1100 : JO=DJ
6815  IF ER(1)=1 THEN RETURN
6820  MN=MO : DY=DO : FL(1)=0 : GOSUB 1100
6825  IF ER(1)=1 THEN RETURN
6830  K=FNL(((YO-1900.0+((DJ-JO)/365.0))*12.3685)+0.5)
6835  TN=K/1236.85 : TF=(K+0.5)/1236.85
6840  T=TN : GOSUB 6855 : NI=A : NF=B : NB=F
6845  T=TF : K=K+0.5 : GOSUB 6855 : FI=A : FF=B : FB=F
6850  RETURN
```

6855 The local routine ...

```
6855  T2=T*T : E=29.53*K : C=166.56+(132.87-9.173E-3*T)*T
6860  C=FNM(C) : B=5.8868E-4*K+(1.178E-4-1.55E-7*T)*T2
6865  B=B+3.3E-4*SIN(C)+7.5933E-1 : A=K/1.236886E1
```

6870 A1 is Sun's mean anomaly
6875 A2 is Moon's mean anomaly

```
6870  A1=359.2242+360*FNW(A)-(3.33E-5+3.47E-6*T)*T2
6875  A2=306.0253+360*FNW(K/9.330851E-1)
6880  A2=A2+(1.07306E-2+1.236E-5*T)*T2 : A=K/9.214926E-1
```

6885 F is Moon's argument of latitude
6890 ... all reduced to fundamental interval 0–360 ...

```
6885  F=21.2964+360.0*FNW(A)-(1.6528E-3+2.39E-6*T)*T2
6890  A1=FNV(A1) : A2=FNV(A2) : F=FNV(F)
```

6895 ... and converted to radians

```
6895  A1=FNM(A1) : A2=FNM(A2) : F=FNM(F)
```

6900 Now calculate the increment of the Julian date

```
6900  DD=(1.734E-1-3.93E-4*T)*SIN(A1)+2.1E-3*SIN(2*A1)
6905  DD=DD-4.068E-1*SIN(A2)+1.61E-2*SIN(2*A2)-4E-4*SIN(3*A2)
6910  DD=DD+1.04E-2*SIN(2*F)-5.1E-3*SIN(A1+A2)
6915  DD=DD-7.4E-3*SIN(A1-A2)+4E-4*SIN(2*F+A1)
6920  DD=DD-4E-4*SIN(2*F+A2)-6E-4*SIN(2*F+A2)+1E-3*SIN(2*F-A2)
6925  DD=DD+5E-4*SIN(A1+2*A2) : E1=INT(E) : B=B+DD+(E-E1)
6930  B1=INT(B) : A=E1+B1 : B=B-B1
```

6935 ... and return from the local routine

```
6935  RETURN
```

```
1      REM
2      REM        Handling program HMOONNF
3      REM

5      DIM FL(20),ER(20),SW(20)
10     DEF FNI(W)=SGN(W)*INT(ABS(W))

15     DEF FNL(W)=FNI(W)+FNI((SGN(W)-1.0)/2.0)
20     DEF FNM(W)=1.745329252E-2*W
25     DEF FND(W)=5.729577951E1*W
30     DEF FNV(W)=W-FNL(W/360.0)*360.0
35     DEF FNW(W)=W-FNL(W)

36     INPUT "^P" ; X
40     PRINT : PRINT : PRINT "New and full moon"
45     PRINT "------------------" : PRINT

50     Q$="Approximate date (D,M,Y)...... "
55     PRINT Q$; : INPUT DY,MN,YR

60     PRINT : GOSUB 6800 : X=NF-0.5 : DJ=NI+0.5

65     IF ER(1)=1 THEN GOTO 145
70     IF X<0 THEN X=X+1 : DJ=DJ-1
75     FL(2)=0 : GOSUB 1200
80     Q$="New moon on date ............ "
85     PRINT Q$+DT$
90     X=X*24.0+8.333E-3 : SW(1)=1 : GOSUB 1000
95     Q$="...and at time (H,M) ........ "
100    PRINT Q$+" "+MID$(OP$,4,6)

105    X=FF-0.5 : DJ=FI+0.5
110    IF X<0 THEN X=X+1 : DJ=DJ-1
115    FL(2)=0 : GOSUB 1200
120    Q$="Full moon on date ........... "
125    PRINT Q$+DT$
130    X=X*24+8.333E-3 : GOSUB 1000
135    Q$="...and at time (H,M) ........ "
140    PRINT Q$+" "+MID$(OP$,4,6)

145    Q$="Again (Y or N) .............. "
150    PRINT : GOSUB 960
155    IF E=0 THEN STOP
160    PRINT : GOTO 50

INCLUDE YESNO, MINSEC, JULDAY, CALDAY, MOONNF
```

Notes (left column):

10 FNI is truncated-integer function (JULDAY)...
15 ... and FNL the least-integer function
20 Converts degrees to radians...
25 ... and radians to degrees
30 FNV restores to the range 0–360 degrees
35 FNW returns the fractional part

50 Get a new (approximate) date...

60 ... and call MOONNF; adjust days for Julian 0.5 day offset
65 Was date acceptable?
70 Adjust for date change...
75 ... and call CALDAY to find calendar date
85 Display the results...
90 ... times in H,M,S format rounded to nearest minute
100 ... neatly

105 Display results also for full moon

145 Another go?
150 Ask YESNO

6800 MOONNF

Example

```
New and full moon
-----------------

Approximate date (D,M,Y)...... ? 12,12,1968

New moon on date .............    19 12  1968
...and at time (H,M) .........    18 19
Full moon on date ............     3  1  1969
...and at time (H,M) .........    18 29

Again (Y or N) ............... ? Y

Approximate date (D,M,Y)...... ? 1,4,1974

New moon on date .............    23  3  1974
...and at time (H,M) .........    21 25
Full moon on date ............     6  4  1974
...and at time (H,M) .........    21  0

Again (Y or N) ............... ? Y

Approximate date (D,M,Y)...... ? 1,9,1984

New moon on date .............    26  8  1984
...and at time (H,M) .........    19 27
Full moon on date ............    10  9  1984
...and at time (H,M) .........     7  3

Again (Y or N) ............... ? N
```

The Astronomical Ephemeris/Almanac gives the following values:

Date	New moon	Full moon
19th December 1968	18 19	
3rd January 1969		18 28
23rd March 1974	21 24	
6th April 1974		21 00
26th August 1984	19 26	
10th September 1984		07 01

7000 ECLIPSE

This routine calculates the circumstances of lunar and solar eclipses.

An eclipse occurs whenever the Moon's apparent position coincides with that of the Sun (solar eclipse) or the Earth's shadow (lunar eclipse). In the former case, the Moon is seen to pass in front of the Sun and obscure it. In the latter case, the bright Moon fades as it passes into the shadow of the Earth, until it is dimly illuminated only by the reddened light refracted through the Earth's atmosphere if the eclipse is total. A solar eclipse can happen only at new moon and a lunar eclipse at full moon. There is not an eclipse on every such occasion, however, since the Moon's orbit is inclined at an angle of about 5 degrees to the plane of the ecliptic. This means that the Moon must also be within about 18.5 degrees of one of its nodes, limiting the total number of eclipses possible each year to a maximum of seven.

Subroutine ECLIPSE (7000) calculates all the circumstances of a lunar or solar eclipse occurring near a given date. This date is input as usual via DY, MN, YR, and the type of eclipse via the string variable ET\$, set to 'L' for a lunar eclipse and to 'S' for a solar eclipse. The observer's position on the Earth is also input via the variables GL (geographical longitude), GP (geographical latitude) and HT (height above sea-level). The routine first finds the times of new and full moon nearest to the given date and then tests to see whether an eclipse is possible. If it is, then the Universal Time (UT) of maximum eclipse is calculated, Z1, together with the times of first contact, Z6, and last contact, Z7, and the eclipse magnitude, MG. If the eclipse is lunar, the start of the umbral phase, Z8, the end of the umbral phase, Z9, the start of the total phase ZC, and the end of the total phase, ZB, are also calculated as appropriate. The magnitude is defined as the fraction of the diameter of the disk (Sun, penumbra, or umbra) covered by the Moon at maximum eclipse, and in the case of a lunar eclipse is the penumbral magnitude if there is no umbral phase.

Error conditions are reported via the error flags ER(8) and ER(9). Both are set to 0 on entering the routine (line 7000). ER(8) is set to 1 if no eclipse occurs at all, and is set to 2 if a solar eclipse occurs somewhere on the Earth but not at

7000 ECLIPSE

the given geographical location. ER(9) is set to 1 if there is no umbral phase in a lunar eclipse, and to 2 if there is no total phase.

On calling the routine, it always responds with one of the messages '*Solar eclipse certain*', '*Lunar eclipse certain*', '*Solar eclipse possible*', '*Lunar eclipse possible*', or '*No eclipse*' referring to the instant of new or full moon tested. This date is returned via the variables D0, MN, YR, and in string format via the variable DT$. Hence it is useful to make the calling program respond with the message '*on*' followed by the date D0, MN, YR or DT$ as does the handling program HECLIPSE. Other messages put out by the routine are '*but does not occur*' when circumstances contrive to prevent an eclipse from occurring despite the Moon being within 18.5 degrees of a node, and '*but is not seen from this locasion*' in the case of a solar eclipse which happens elsewhere on the Earth. Note that ECLIPSE does *not* test to see whether the Moon or Sun is above the observer's horizon during the eclipse (but DISPLAY (500) can do so; see later in this section). Hence the times given are appropriate to a transparent Earth and you must use the routines SUNRS (3600) and MOONRS (6600) to find whether the eclipse can actually be observed from the given location. These routines could be incorporated into HECLIPSE so that the testing is carried out automatically, but I felt that the code would then become rather cumbersome and slow, and I have therefore not attempted it myself.

If your machine has the capability of drawing pictures, you can display the eclipse on the monitor screen. Line 315 of the handling program allows this option by calling a display subroutine at line 500. If you cannot, or do not wish to use the graphics display, simply put a RETURN statement at line 500. Otherwise, you can write your own routine here. Unfortunately, the exact form of the routine will depend on which machine you are using so I have not been able to supply universal code. However, I have included some notes on how to proceed, together with the listing of the routine that produced the eclipse pictures printed here which you can adapt for your own machine as you see fit. (See 'eclipse graphics' later in this section.)

Formulae
See Figure 6 for definitions of some of the quantities.

$ZP = Z^3/(Z^2 + S^2)$

$ZQ = ZP - (ZP^2 - ((Z^2 - (R1 + RM)^2) \times ZP/Z))^{0.5}$

RM = radius of Moon = $2.72446E{-}1 \times PM$

PM = Moon's horizontal parallax (radians)

R1 = radius of Sun (RN), radius of penumbra (RP), or radius of umbra (RU)

RN = $4.65242E{-}3/RR$ (radians)

RR = Earth–Sun distance (AU)

178

RP = ((PM×0.99834) + RN + PS) × 1.02
RU = ((PM×0.99834) − RN + PS) × 1.02
PS = parallax of Sun
MG = (RM + R1 − P)/(2×R1) for solar eclipse with R1 = RN
MG = (RM + R1 − P)/(2×RM) for lunar eclipse with R1 = RP for
        penumbral eclipse and R1 = RU for umbral eclipse
P = Z×S/(Z² + S²)^0.5

Details of ECLIPSE

Called by GOSUB 7000.

Calculates Z1, the UT of maximum eclipse, Z6, the UT of first contact, Z7, the UT of last contact, Z8, the UT of the start of the umbral phase (lunar), Z9, the UT of the end of the umbral phase, ZC, the UT of the start of the total phase (lunar), and ZB, the UT of the end of the total phase. All times in hours. The routine also returns with the date of the new or full moon nearest to the input date (and hence the date of the eclipse if there is one) via D0 (days), MN (months), YR (years) and DT$ (in string format), the eclipse magnitude MG, and various other parameters used in the graphics routine (see below). Input parameters are DY (days), MN (months), YR (years), the approximate date, ET$, the eclipse type ('L' for lunar and 'S' for solar), GL (*degrees*), the geographical longitude (West negative), GP (*radians*), the geographical latitude, and HT (units of Earth radii), the height above sea-level.

ER(8) is set to 1 if there is no eclipse, and to 2 if a solar eclipse occurs elsewhere on the Earth, but not at the given location. Otherwise ER(8) = 0.

ER(9) is set to 1 when there is a lunar eclipse without an umbral phase, and to 2 when there is an umbral phase but no total phase. Otherwise ER(9) = 0.

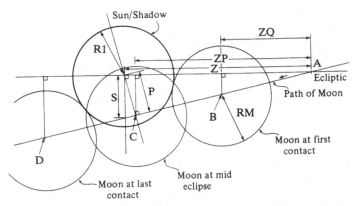

Figure 6. Eclipse geometry, defining some of the variables used in the subroutine ECLIPSE.

7000 ECLIPSE

Other routines called: CALDAY (1200), TIME (1300), HRANG (1600), NUTAT (1800), EQECL (2000), PARALLX (2800), SUN (3400), MOON (6000), and MOONNF (6800).

```
6997 REM
6998 REM        Subroutine ECLIPSE
6999 REM
```

7000	Reset error flags, and define constants
7005	Call MOONNF to find dates of new and moon
7010	Set DF equal to the absolute distance . . .
7015	. . . from the Moon's node . . .
7020	. . . taking care of DF + 180 degrees
7025	Use DF to test for eclipse . . .
7030	. . . spot on . . .
7035	. . . or maybe . . .
7045	. . . or definitely not
7050	Set Julian date DJ of the event, either new . . .
7055	. . . or full moon . . .
7060	. . . adjusting for change of day if needed
7065	Call CALDAY to get calendar date of event . . .
7070	. . . and return if there is no eclipse
7075	Call SUN for position of Sun 1 hour before event . . .
7080	Surely the date could never be impossible?
7085	. . . and MOON for the Moon's position . . .
7090	. . . and repeat for both Sun . . .
7100	. . . and Moon for 1 hour after the event.
7105	Skip the next bit if it's a lunar eclipse
7110	Overkill with the flags; call NUTAT for nutation
7120	Call local parallax routine . . .
7125	. . . to adjust Moon's apparent positions . . .
7130	. . . for the observer's location
7135	X0 is the time of zero lunar latitude, marked A in Figure 6
7140	DM is difference in lunar longitudes
7145	LJ is longitude difference to hours difference scale factor
7150	MR is longitude of Moon at A
7155	Allow for light-travel time from Sun, call SUN, . . .
7160	. . . and allow for aberration in solar longitude
7170	If lunar eclipse, we want longitude of shadow . . .
7175	If solar eclipse, call local parallax routine . . .
7180	. . . to find apparent position of the Sun
7185	Make corrections for parallax in latitude

```
7000 ER(8)=0 : ER(9)=0 : PI=3.141592654 : TP=2.0*PI
7005 LE$="Lunar " : SE$="Solar " : GOSUB 6800

7010 IF ET$="L" THEN DF=ABS(FB-FNL(FB/PI)*PI) : Q$=LE$
7015 IF ET$="S" THEN DF=ABS(NB-FNL(NB/PI)*PI) : Q$=SE$
7020 IF DF>0.37 THEN DF=PI-DF

7025 IF DF>=2.42600766E-1 THEN GOTO 7035
7030 PRINT Q$+"eclipse certain "; : GOTO 7050

7035 IF DF>=0.37 THEN GOTO 7045
7040 PRINT Q$+"eclipse possible "; : GOTO 7050

7045 PRINT "No eclipse "; : ER(8)=1

7050 DJ=NI+0.5 : X=NF-0.5 : DP=0

7055 IF ET$="L" THEN DJ=FI+0.5 : X=FF-0.5
7060 IF X<0 THEN X=X+1 : DJ=DJ-1
7065 FL(2)=0 : GOSUB 1200 : D0=DY

7070 IF ER(8) <> 0 THEN RETURN

7075 UT=24.0*X-1.0 : XI=X : FL(1)=0 : GOSUB 3400

7080 IF ER(1)<>0 THEN RETURN

7085 LY=SR : GOSUB 6000 : MY=MM : BY=BM : HY=PM

7090 UT=UT+2.0 : FL(1)=0 : GOSUB 3400 : SB=SR-LY
7095 IF SB<0 THEN SB=SB+TP
7100 GOSUB 6000 : MZ=MM : BZ=BM : HZ=PM : XH=XI*24.0
7105 IF ET$="L" THEN GOTO 7135

7110 FL(5)=0 : FL(6)=0 : FL(7)=0 : GOSUB 1800
7115 X=MY : Y=BY : TM=XH-1.0 : HP=HY
7120 GOSUB 7390 : MY=P : BY=Q
7125 X=MZ : Y=BZ : TM=XH+1.0 : HP=HZ

7130 GOSUB 7390 : MZ=P : BZ=Q

7135 X0=XH+1.0-(2.0*BZ/(BZ-BY)) : DM=MZ-MY

7140 IF DM<0 THEN DM=DM+TP
7145 LJ=(DM-SB)/2.0 : Q=0

7150 MR=MY+(DM*(X0-XH+1.0)/2.0)
7155 UT=X0-1.3851852E-1 : FL(1)=0 : GOSUB 3400

7160 SR=SR+FNM(DP-5.69E-3)
7165 IF ET$="S" THEN GOTO 7175
7170 SR=SR+PI-FNL((SR+PI)/TP)*TP : GOTO 7185

7175 X=SR : Y=0 : TM=UT

7180 HP=4.263452E-5/RR : GOSUB 7390 : SR=P

7185 BY=BY-Q : BZ=BZ-Q : P3=4.263E-5
```

7000 ECLIPSE

7190	Now calculate geometry of eclipse (Figure 6)	`7190 ZH=(SR-MR)/LJ : TC=X0+ZH`
		`7195 SH=(((BZ-BY)*(TC-XH-1.0)/2.0)+BZ)/LJ`
		`7200 S2=SH*SH : Z2=ZH*ZH : PS=P3/(RR*LJ)`
7205	Z1 is UT of maximum eclipse	`7205 Z1=(ZH*Z2/(Z2+S2))+X0`
7210	Use average parallax and find radius of Moon RM . . .	`7210 H0=(HY+HZ)/(2.0*LJ) : RM=2.72446E-1*H0`
7215	. . . radius of Sun RN . . .	`7215 RN=4.65242E-3/(LJ*RR) : HD=H0*0.99834`
7220	. . . radius of umbra RU, and radius of penumbra, RP	`7220 RU=(HD-RN+PS)*1.02 : RP=(HD+RN+PS)*1.02`
		`7225 PJ=ABS(SH*ZH/SQR(S2+Z2))`
7230	Skip this bit if lunar eclipse	`7230 IF ET$="L" THEN GOTO 7280`
7250	Solar eclipse only . . .	`7250 R=RM+RN : DD=Z1-X0 : DD=DD*DD-((Z2-(R*R))*DD/ZH)`
7255	. . . can it be seen from here? Abort if not	`7255 IF DD<0 THEN GOTO 7370`
7260	Z6 is UT of first contact . . .	`7260 ZD=SQR(DD) : Z6=Z1-ZD`
7265	. . . and Z7 the UT of last contact	`7265 Z7=Z1+ZD-FNL((Z1+ZD)/24.0)*24.0`
7270	Make sure Z6 is in range 0–24	`7270 IF Z6<0 THEN Z6=Z6+24.0`
7275	Calculate magnitude, MG, and return from ECLIPSE	`7275 MG=(RM+RN-PJ)/(2.0*RN) : RETURN`
7280	Lunar eclipse only; calculate penumbral phase . . .	`7280 R=RM+RP : DD=Z1-X0 : DD=DD*DD-((Z2-(R*R))*DD/ZH)`
7285	. . . aborting if there is not one . . .	`7285 IF DD<0 THEN GOTO 7380`
7290	Z6 is UT of first contact . . .	`7290 ZD=SQR(DD) : Z6=Z1-ZD`
7295	. . . and Z7 is the UT of last contact	`7295 Z7=Z1+ZD-FNL((Z1+ZD)/24.0)*24.0`
7300	Make sure Z6 is in range 0–24	`7300 IF Z6<0 THEN Z6=Z6+24.0`
		`7305 R=RM+RU : DD=Z1-X0 : DD=DD*DD-((Z2-(R*R))*DD/ZH)`
7310	Calculate penumbral magnitude . . .	`7310 MG=(RM+RP-PJ)/(2.0*RM)`
7315	. . . and return from ECLIPSE if no umbral phase	`7315 IF DD<0 THEN ER(9)=1 : RETURN`
7320	Calculate umbral phase; Z8 is UT of start . . .	`7320 ZD=SQR(DD) : Z8=Z1-ZD`
7325	. . . and Z9 is UT of end	`7325 Z9=Z1+ZD-FNL((Z1+ZD)/24.0)*24.0`
7330	Make sure that Z8 is in range 0–24	`7330 IF Z8<0 THEN Z8=Z8+24.0`
		`7335 R=RU-RM : DD=Z1-X0 : DD=DD*DD-((Z2-(R*R))*DD/ZH)`
7340	Calculate umbral magnitude . . .	`7340 MG=(RM+RU-PJ)/(2.0*RM)`
7345	. . . and return from ECLIPSE if no total phase	`7345 IF DD<0 THEN ER(9)=2 : RETURN`
7350	Calculate total phase; ZC is UT of start . . .	`7350 ZD=SQR(DD) : ZC=Z1-ZD`
7355	. . . and ZB is UT of end	`7355 ZB=Z1+ZD-FNL((Z1+ZD)/24.0)*24.0`
7360	Make sure ZC is in range 0–24	`7360 IF ZC<0 THEN ZC=ZC+24.0`
7365	Return from ECLIPSE	`7365 RETURN`
7370	Return with error messages	`7370 PRINT : PRINT "but is not seen from this location "`
		`7375 ER(8)=2 : RETURN`
		`7380 PRINT : PRINT "but does not occur ";`
		`7385 ER(8)=1 : RETURN`
7390	Local parallax routine; call EQECL for right ascension and declination . . .	`7390 SW(3)=-1 : GOSUB 2000 : CN=3.819718634`
7395	. . . call TIME for local sidereal time . . .	`7395 X1=P*CN : SW(2)=1 : FL(3)=0 : GOSUB 1300`
7400	. . . call HRANG for conversion from right ascension to hour angle . . .	`7400 X=X1 : GOSUB 1600 : SW(5)=1`
7405	. . . call PARALLX to correct hour angle and declination . . .	`7405 X=P/CN : Y=Q : GOSUB 2800 : Y=Q : X=P*CN`
7410	. . . call HRANG again to convert hour angle to right ascension, and then EQECL . . .	`7410 GOSUB 1600 : X=P/CN : SW(3)=1 : GOSUB 2000`
7415	. . . to convert back to ecliptic coordinates	`7415 RETURN`

```
1     REM
2     REM         Handling program HECLIPSE
3     REM

5     DIM FL(20),ER(20),SW(20),CC(3),CV(3),HL(3),MT(3,3)
```

10 FNI is truncated-integer function . . .
```
10    DEF FNI(W)=SGN(W)*INT(ABS(W))
```
15 . . . and FNL is least-integer function
```
15    DEF FNL(W)=FNI(W)+FNI((SGN(W)-1.0)/2.0)
```
20 FNM converts degrees to radians . . .
```
20    DEF FNM(W)=1.745329252E-2*W
```
25 . . . and FND converts radians to degrees
```
25    DEF FND(W)=5.729577951E1*W
```
30 FNS returns the inverse sine
```
30    DEF FNS(W)=ATN(W/(SQR(1-W*W)+1E-20))
```
35 FNV restores to the range 0–360
```
35    DEF FNV(W)=W-FNL(W/360.0)*360.0
```
40 FNW returns the fractional part
```
40    DEF FNW(W)=W-FNL(W)

45    PRINT : PRINT : PRINT "Lunar and solar eclipses"
50    PRINT "----------------------" : PRINT : SW=0
55    PRINT : SW(1)=-1 : TZ=0 : DS=0
```

60 FL(12)=1 indicates that the location has not changed
```
60    IF FL(12)=1 THEN GOTO 100
```
65 Get details of a new location . . .
```
65    Q$="Geographical longitude (W neg; D,M,S) .. "
```
70 . . . longitude GL in degrees (via MINSEC)
```
70    PRINT Q$; : INPUT XD,XM,XS : GOSUB 1000 : GL=X
75    Q$="Geographical latitude (D,M,S) .......... "
```
80 . . . latitude GP in radians
```
80    PRINT Q$; : INPUT XD,XM,XS : GOSUB 1000 : GP=FNM(X)
85    Q$="Height above sea-level (m) ............ "
```
90 . . . and height above sea-level HT in Earth radii
```
90    PRINT Q$; : INPUT X : HT=X/6378140.0
```
95 SW is a local switch meaning 'same dates' if set to 1
```
95    IF SW=1 THEN PRINT : GOTO 120
```
100 Get new date . . .
```
100   Q$="Starting date (D,M,Y) ................. "
105   PRINT Q$; : INPUT DY,MN,YR
110   Q$="For how many months .................. "
```
115 . . . and length of search window
```
115   PRINT Q$; : INPUT FH : PRINT
```

120 KO is a local counter
```
120   IF SW=0 THEN ET$="S" : KO=KO+1
```
125 Call ECLIPSE
```
125   FL(1)=0 : FL(3)=0 : FL(5)=0 : GOSUB 7000 : SW=0
```
130 Impossible date (should never happen)
```
130   IF ER(1) <> 0 THEN GOTO 445
```
135 ER(8) indicates whether there was an eclipse . . .
```
135   IF ER(8)<>0 THEN Q$="on" : PRINT Q$+DT$;
140   IF ER(8)=2 THEN PRINT : PRINT : GOTO 400
145   IF ER(8) <> 0 THEN GOTO 420
```

150 . . . ER(8)=0 means there was one
```
150   IF ET$="L" THEN Q$="Lunar eclipse of"
155   IF ET$="S" THEN Q$="Solar eclipse of"
```
160 Announce its type and date
```
160   PRINT : PRINT : PRINT TAB(10); Q$+DT$
165   PRINT TAB(10); "------------------------------"
```
170 Round all times to nearest minute . . .
```
170   X=Z1+8.33E-3 : X=X-FNL(X/24)*24
```
175 . . . convert to H,M,S format . . .
```
175   SW(1)=1 : GOSUB 1000 : PRINT : U1$=OP$
180   Q$="UT of maximum eclipse ................. "
```
185 . . . and display the results neatly
```
185   PRINT Q$+" "+MID$(OP$,4,6)
190   X=Z6+8.33E-3 : X=X-FNL(X/24)*24 : GOSUB 1000
195   Q$="Eclipse begins at ..................... "
200   U2$=OP$ : PRINT Q$+" "+MID$(OP$,4,6)
205   X=Z7+8.33E-3 : X=X-FNL(X/24)*24 : GOSUB 1000
210   Q$="Eclipse ends at ....................... "
215   U3$=OP$ : PRINT Q$+" "+MID$(OP$,4,6)
```
220 That's all for solar eclipses
```
220   IF ET$="S" THEN GOTO 305
```

225 . . . and for lunar eclipse if no umbral phase
```
225   IF ER(9) <> 1 THEN GOTO 235
230   PRINT "** no umbral phase **" : GOTO 305
```

235 Umbral phase: display the results
```
235   X=Z8+8.33E-3 : X=X-FNL(X/24)*24 : GOSUB 1000
240   Q$="Umbral phase begins at ................ "
245   U4$=OP$ : PRINT Q$+" "+MID$(OP$,4,6)
250   X=Z9+8.33E-3 : X=X-FNL(X/24)*24 : GOSUB 1000
```

	255	Q$="Umbral phase ends at .... "
	260	U7$=OP$: PRINT Q$+" "+MID$(OP$,4,6)
265	That's all, unless . . .	265	IF ER(9)<>2 THEN GOTO 275
		270	PRINT "** no total phase **" : GOTO 305
275	. . . there's a total phase; display the results	275	X=ZC+8.33E-3 : X=X-FNL(X/24)*24 : GOSUB 1000
		280	Q$="Total phase begins at .... "
		285	U5$=OP$: PRINT Q$+" "+MID$(OP$,4,6)
		290	X=ZB+8.33E-3 : X=X-FNL(X/24)*24 : GOSUB 1000
		295	Q$="Total phase ends at .... "
		300	U6$=OP$: PRINT Q$+" "+MID$(OP$,4,6)
305	Always report the magnitude . . .	305	Q$="Eclipse magnitude .... "
310	. . . to three decimal places	310	PRINT Q$+" "+STR$(INT((MG*1000)+0.5)/1000)
315	Have you written a display routine?	315	Q$="Display the eclipse (Y or N) .... "
320	Ask YESNO to get the answer	320	PRINT : GOSUB 960
325	Skip to end if not	325	IF E=0 THEN GOTO 400
330	What part of the eclipse do you wish to display?	330	Q$="First contact (Y or N) .... "
332	ZJ and UE$ are the input parameters to DISPLAY . . .	332	GOSUB 960 : ZJ=Z6 : UE$=U2$
334	. . . which begins at line 500	334	IF E=1 THEN GOSUB 500 : GOTO 385
		336	IF ET$="S" OR ER(9)=1 THEN GOTO 352
		338	Q$="Start of umbral phase (Y or N) .... "
		340	GOSUB 960 : ZJ=Z8 : UE$=U4$
		342	IF E=1 THEN GOSUB 500 : GOTO 385
		344	IF ER(9)=2 THEN GOTO 352
		346	Q$="Start of total phase (Y or N) .... "
		348	GOSUB 960 : ZJ=ZC : UE$=U5$
		350	IF E=1 THEN GOSUB 500 : GOTO 385
		352	Q$="Mid eclipse (Y or N) .... "
		354	GOSUB 960 : ZJ=Z1 : UE$=U1$
		356	IF E=1 THEN GOSUB 500 : GOTO 385
		358	IF ET$="S" OR ER(9)>0 THEN GOTO 366
		360	Q$="End of total phase (Y or N) .... "
		362	GOSUB 960 : ZJ=ZB : UE$=U6$
		364	IF E=1 THEN GOSUB 500 : GOTO 385
		366	IF ET$="S" OR ER(9)=1 THEN GOTO 374
		368	Q$="End of umbral phase (Y or N) .... "
		370	GOSUB 960 : ZJ=Z9 : UE$=U7$
		372	IF E=1 THEN GOSUB 500 : GOTO 385
		374	Q$="Last contact (Y or N) .... "
		376	GOSUB 960 : ZJ=Z7 : UE$=U3$
		378	IF E=1 THEN GOSUB 500 : GOTO 385
		380	GOTO 315
385	Display another aspect of the eclipse?	385	Q$="Display again (Y or N) ...."
390	YESNO returns the answer	390	GOSUB 960
		395	IF E=1 THEN GOTO 330
400	Calculate the solar eclipse at another place?	400	IF ET$="L" THEN GOTO 420
		405	Q$="Another location (Y or N) .... "
		410	GOSUB 960
415	If yes, reset the switches and jump back	415	IF E=1 THEN SW=1 : FL(12)=0 : GOTO 55
420	Search for the lunar eclipse near this date	420	IF ET$="S" THEN ET$="L" : PRINT : GOTO 125
425	Increment current date by 14 days and call JULDAY	425	DY=D0+14 : FL(1)=0 : GOSUB 1100
430	We might now have an impossible date; ask for help if so	430	IF ER(1) <> 0 THEN GOTO 445
435	. . . Convert to calendar date . . .	435	FL(2)=0 : GOSUB 1200
440	. . . and go round again, unless we have done enough	440	IF KO<FH THEN PRINT : GOTO 120

445 Another go?
450 Get the answer with YESNO

460 Change the default location?
465 Reset FL(12) to 0 for a new place

```
445  Q$="Again (Y or N) ......................... "
450  PRINT : PRINT : GOSUB 960
455  IF E=0 THEN STOP
460  Q$="Same location (Y or N) ................. "
465  GOSUB 960 : FL(12)=E : GOTO 55

INCLUDE DISPLAY, YESNO, MINSEC, JULDAY, CALDAY, TIME,
        HRANG, OBLIQ, NUTAT, EQECL, GENCON, PARALLX,
        ANOMALY, SUN, MOON, MOONNF, ECLIPSE
```

7000 ECLIPSE

Example

```
Lunar and solar eclipses
------------------------

Geographical longitude (W neg; D,M,S) .. ? -5,45,0
Geographical latitude (D,M,S) .......... ? 50,0,0
Height above sea-level (m) ............. ? 0
Starting date (D,M,Y) .................. ? 1,7,1999
For how many months ................... ? 3

No eclipse on  13  7  1999
Lunar eclipse certain

          Lunar eclipse of  28  7  1999
          ----------------------------

UT of maximum eclipse .................    11 34
Eclipse begins at .....................     8 57
Eclipse ends at .......................    14 12
Umbral phase begins at ................    10 22
Umbral phase ends at ..................    12 46
** no total phase **
Eclipse magnitude .....................     0.402

Display the eclipse (Y or N) .......... ? N

Solar eclipse certain

          Solar eclipse of  11  8  1999
          ----------------------------

UT of maximum eclipse .................    10 11
Eclipse begins at .....................     8 52
Eclipse ends at .......................    11 31
Eclipse magnitude .....................     1.004

Display the eclipse (Y or N) .......... ? N
Another location (Y or N) ............. ? Y

Geographical longitude (W neg; D,M,S) .. ? 18,0,0
Geographical latitude (D,M,S) .......... ? 42,30,0
Height above sea-level (m) ............. ? 0

Solar eclipse certain

          Solar eclipse of  11  8  1999
          ----------------------------

UT of maximum eclipse .................    10 53
Eclipse begins at .....................     9 28
Eclipse ends at .......................    12 18
Eclipse magnitude .....................     0.895

Display the eclipse (Y or N) .......... ? N
Another location (Y or N) ............. ? N

No eclipse on  26  8  1999
No eclipse on   9  9  1999
No eclipse on  25  9  1999

Again (Y or N) ........................ ? N
```

Tests against the values quoted in *The Astronomical Almanac* and *The Astronomical Ephemeris* show that the times are generally correct to within a minute or two provided that machine rounding errors are insignificant.

ECLIPSE GRAPHICS – DISPLAY

The eclipse pictures printed in Figure 7, which show lunar and solar eclipses at various stages, were produced by subroutine DISPLAY listed later in this section. Unfortunately, it is not possible to write code which will work equally well on every computer since the mechanics of producing graphics displays varies so much from machine to machine. You will certainly have to modify DISPLAY to suit the particular characteristics of your own computer. To help you do so, I have collected all the graphics commands into a number of function calls; for example, FNCLGS clears the graphics screen to black. You can use DISPLAY just as it is provided you can define the functions so that they work on your machine. Alternatively, you may replace the functions with calls to procedures or subroutines to do the same thing. Thus, if your BASIC includes the graphics command 'HOME', for example, which clears the screen, you could substitute 'GOSUB 9000' for 'X=FNCLGS' and include the extra sub-routine

9000 HOME : RETURN

at the end. Obviously, more complex procedures will require several lines of code each.

The graphics screen on my computer measures 392 pixels in the horizontal, or X direction, and 256 in the vertical, or Y direction. You can think of the screen as a large sheet of graph paper, with the point X=0, Y=0 at the bottom left-hand corner of the screen. Positions are usually referred to in the order X,Y. To place a dot in the centre of the screen I would need to light up the pixel at point (196,128). Your screen probably has different dimensions, and so you will need to alter some of the numbers in DISPLAY as well as redefining the graphics function calls. The line-by-line notes on the left-hand page should help you do so.

Routine DISPLAY includes the option of displaying the eclipse relative to a horizontal ecliptic, or rotated so that it appears just as it would when viewed from the ground. Lines 515–552 calculate the value of RO (degrees) which is the rotation angle anticlockwise. If you cannot incorporate rotation into your

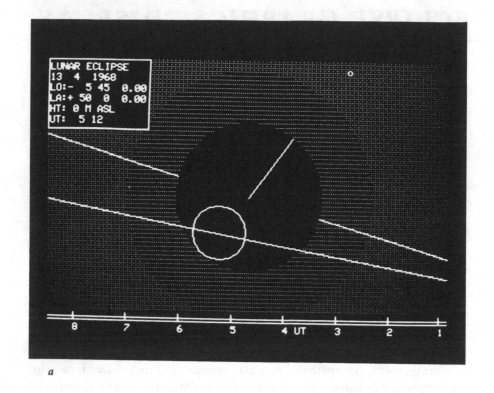

a

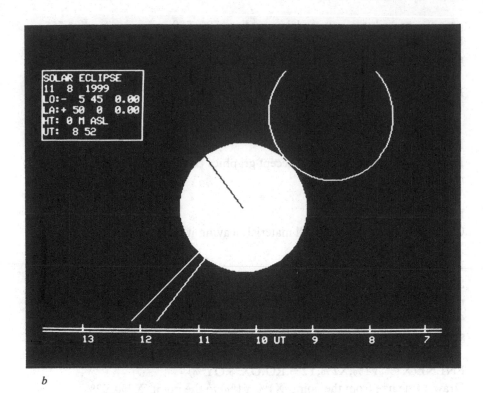

SOLAR ECLIPSE
11 8 1999
LO:- 5 45 0.00
LA:+ 50 0 0.00
HT: 0 M ASL
UT: 8 52

13   12   11   10 UT   9   8   7

b

Figure 7. Eclipse graphics. These diagrams are black and white reproductions of the colour monitor screen produced by the subroutine DISPLAY.

The lunar eclipse of 13 April 1968 is shown in (a) at the last moment of the total phase. The dark inner circle is the Earth's umbra, and the lighter outer circle is the penumbra. The small unfilled circle is the Moon, and the sloping line through it is its path relative to the centre of the umbra. The other sloping line, which passes through the centre of the umbra, represents the plane of the ecliptic. The short line from the centre of the umbra upwards towards the right is the direction of North, i.e. the trace of the great circle through the North Celestial Pole and the centre of the umbra.

The total solar eclipse of 11 August 1999 is shown in (b) at the moment of first contact. The white circle in the centre is the Sun, with the direction of North shown by the sloping line up towards the left. The unfilled circle at the top right of the diagram is the Moon. During the eclipse, the Moon passes in front of the Sun along the path indicated by the upper of the two lines sloping down towards the left. The other line shows the plane of the ecliptic, i.e. the path of the Sun.

Both diagrams have been rotated to represent the view as seen from the ground. The horizontal time scales, which show the position of the Moon at each hour, are therefore parallel to the horizon.

500 DISPLAY

graphics procedures, you can leave out lines 500–550 and renumber line 555 to become 500. You must also leave out the references to RO elsewhere in the routine, and set RW=0 to bypass lines 710 and 715.

Details of the graphics procedures

The % symbol following a variable indicates that the variable is an integer.

FNINIT

Initialises the graphics card to accept graphics commands. You may not need this procedure.

FNCLGS

Clear the graphics screen of all material, leaving it black.

FNCOL(I%)

Set the colour to be used in all following graphics commands until explicitly reset. I% takes the values 0–9 as follows: 0: black, 1: red, 2: green, 3: yellow, 4: blue, 5: magenta, 6: cyan, 7: white, 8: dark grey (almost black), 9: medium grey.

FNLNE(X1%,Y1%,X2%,Y2%,RO,OX%,OY%)

Draw a fine line from the point (X1%,Y1%) to the point (X2%,Y2%), rotated anticlockwise by RO degrees about the point (OX%,OY%). The line is first imagined as being drawn between the given points, then rotated, and then displayed in its rotated state.

FNCIRCLE(X%,Y%,R%, F%,RO,OX%,OY%)

Draw a circle centred on the point (X%,Y%), after the point has been rotated anticlockwise by RO degrees about the point (OX%,OY%). The radius is R% and the circle is unfilled if F%=0, filled in the current colour if F%=1.

FNPLOTM(I%)

Set the method of plotting each pixel according to the value of I%, as follows: 0: overwrite whatever was there before, 1: OR the new value with the existing value, and plot the result, 2: XOR (exclusive OR) the new value with the old before plotting, 3: AND the new value with the old, 4: XNOR (exclusive NOR) the new value with the old, 5: NOT – invert the colour of whatever is already there (e.g. white becomes black etc.), 6: NOR the new value with the old, 7: NAND (NOT AND) the new with the old. The default setting, after FNINIT, is 0.

FNWCOL
Fills in the whole of the current plotting window with the current plotting colour (FNCOL) and the current plotting mode (FNPLOTM).

FNWIND(X1%,X2%,Y1%,Y2%)
Set the current plotting window to the bounds indicated by (X1%,Y1%), bottom left-hand corner, and (X2%,Y2%), top right-hand corner. All plotting commands act normally within the window, and are ignored outside it. The default setting after FNINIT is to allow plotting over the entire graphics screen, i.e. FNWIND(0,391,0,255).

FNTEXT(X%,Y%,C$)
Display the text carried in the string variable C$ beginning at the point (X%,Y%).

FNBOX(X%,Y%,LX%,LY%)
Draw a box with the point (X%,Y%) at the lower left-hand corner with sides of length LX% (horizontal) and LY% (vertical).

FNPRNT
Print out the graphics picture on the dot matrix printer, with each coloured pixel printed black, and each black pixel printed white.

The display algorithm
Initialise the graphics card, and clear the screen to black.
Set colour to green, and draw a line to represent the ecliptic. This should pass through the centre of the screen, and may be rotated, if so desired, to represent the view from the Earth's surface.
LUNAR Set the scale so that the penumbra just fills the screen vertically.
Set colour to white.
Draw a filled circle of the Moon to scale at its (rotated) position corresponding to the event being displayed (e.g. first contact).
Set colour to medium grey.
Set plotting mode to OR.
Colour in the whole screen.
Set colour to dark grey – almost black.
Set plotting mode to XOR.
Draw filled circle of penumbra to scale centred on the screen.
Set colour to black.
Set plotting mode to overwrite whatever was there before.

500 DISPLAY

Draw a filled circle of the umbra to scale centred on the screen.
Set colour to white.
Draw unfilled circle of Moon to scale at (rotated) position of the event (gives its outline).
Draw a straight line to represent the (rotated) path of the Moon.
SOLAR Set scale so that Sun fits screen.
Set colour to yellow.
Draw a filled circle of the Sun to scale centred on the screen.
Set colour to red.
Draw straight line to represent the (rotated) path of the Moon.
Set colour to black.
Draw a filled circle of Moon to scale at the (rotated) position of the event.
Set colour to white.
Draw unfilled circle of Moon at the same position (gives the outline).
Both Set colour to magenta.
Set plotting mode to XOR.
Draw a line from the centre of the screen of length equal to the scaled radius of the umbra (lunar eclipse) or the Sun (solar eclipse) in the direction of the North celestial pole.
Select another colour (e.g. cyan).
Draw a scale along the bottom of the screen to represent the universal times of the events in hours.
Draw in the text at the top left-hand corner of the screen with a box around it.

```
                                    497   REM
                                    498   REM        Subroutine DISPLAY
                                    499   REM
```

500 Draw the ecliptic at its apparent angle to the ground?

```
500   Q$="Rotate ecliptic (Y or N) .............. "
```

505 Get the answer with YESNO

```
505   GOSUB 960 : RO=0 : RW=1
```

510 If it is no, leave RO set to 0 (horizontal ecliptic)

```
510   IF E=0 THEN RW=0 : GOTO 555
```

515 ZJ is Universal Time (UT) of event; call SUN

```
515   DS=0 : TZ=0 : UT=ZJ : GOSUB 3400
```

520 Correct longitude of SUN to get umbra if lunar

```
520   IF ET$="L" THEN SR=SR+PI
```

522 Set up for entry into GENCON . . .

```
522   CC(1)=6 : CC(2)=4 : CC(3)=2 : FL(4)=0 : FL(9)=0
```

524 . . . getting the local sidereal time via TIME . . .

```
524   TM=ZJ : SW(2)=1 : GOSUB 1300 : ST=FNM(TL*15.0)
```

526 . . . to find azimuth and altitude a bit before . . .

```
526   Y=0 : X=SR+0.01 : GOSUB 2200 : P1=P : Q1=Q
```

528 . . . and a bit after the event, to find angle of ecliptic . . .

```
528   X=SR-0.01 : GOSUB 2200 : RO=FND(ATN((Q-Q1)/(P-P1)))
```

530 . . . with respect to the horizon. Declare the results . . .

```
530   AZ=(P+P1)/2.0 : X=FND(AZ) : SW(1)=1 : GOSUB 1000
```

532

```
532   Q$="Azimuth (D,M,S; North = 0) ............ "
```

534 . . . azimuth in D,M,S format . . .

```
534   PRINT : PRINT Q$+OP$
536   X=FND((Q+Q1)/2.0) : GOSUB 1000
538   Q$="Altitude (D,M,S) ...................... "
```

540 . . . and the altitude (You can't see the event if . . .

```
540   PRINT Q$+OP$
```

542 . . . altitude is negative!) . . .

```
542   X=-RO : GOSUB 1000
```

544 . . . and the angle of the ecliptic to the horizon

```
544   Q$="Ecliptic angle from horiz. (D,M,S CW) .. "
```

546 Calculate the direction of North . . .

```
546   PRINT Q$+OP$ : TH=90.0-(FND(GP)-90.0)*SIN(AZ)
548   X=90-TH : GOSUB 1000
```

550 . . . CW means 'clockwise' . . .

```
550   Q$="North angle from vertical (D,M,S CW) ... "
```

552 . . . and display it in D,M,S format

```
552   PRINT Q$+OP$
```

555 Carry event time difference over 24 hour interval

```
555   IF Z1-ZJ>12 THEN ZJ=ZJ+24.0
560   IF ZJ-Z1>12 THEN ZJ=ZJ-24.0
```

565 (OX%,OY%) is centre point of my screen; initialise

```
565   OX%=196 : OY%=128 : X=FNINIT : X=FNCLGS
```

570 Select green and draw line of ecliptic

```
570   X=FNCOL(2) : X=FNLNE(-100,128,500,128,RO,OX%,OY%)
```

575 Skip the next bit if solar eclipse

```
575   IF ET$="S" THEN GOTO 640
```

580 SC is scale factor; R% is scaled radius of Moon

```
580   SC=119/RP : R%=INT%(SC*RM)
```

585 Find the centre point of the Moon . . .

```
585   X%=OX%+INT%((ZH-ZJ+X0)*SC)
590   Y%=OY%+INT%((SH*(ZJ-X0)/ZH)*SC)
```

595 . . . select white, and draw filled circle of Moon

```
595   X=FNCOL(7) : X=FNCIRCLE(X%,Y%,R%,1,RO,OX%,OY%)
```

600 Select medium grey and plot mode OR; fill whole screen

```
600   X=FNCOL(9) : X=FNPLOTM(1) : X=FNWCOL
```

605 R% is scaled radius of penumbra; select dark grey . . .

```
605   R%=INT%(SC*RP) : X=FNCOL(8) : X=FNPLOTM(2)
```

610 . . . and plot mode XOR; draw filled circle of penumbra

```
610   X=FNCIRCLE(OX%,OY%,R%,1,0.0,OX%,OY%)
```

615 R1% is scaled radius of umbra; select black and . . .

```
615   R1%=INT%(SC*RU) : X=FNCOL(0) : X=FNPLOTM(0)
```

620 . . . plot mode 'replace'; draw filled circle

```
620   X=FNCIRCLE(OX%,OY%,R1%,1,0.0,OX%,OY%)
```

193

625	R% is scaled radius of Moon; select white . . .	`625  R%=INT%(SC*RM) : X=FNCOL(7)`
630	. . . draw circle of Moon	`630  X=FNCIRCLE(X%,Y%,R%,0,RO,OX%,OY%)`
635	End of lunar only bit	`635  GOTO 650`
640	Solar only; SC is scale factor; R1% is scaled radius . . .	`640  SC=60.0/RN : R1%=INT%(SC*RN) : X=FNCOL(3)`
645	. . . of Sun; select yellow and plot filled circle of Sun	`645  X=FNCIRCLE(OX%,OY%,R1%,1,0.0,OX%,OY%)`
650	Lunar and solar; calculate path of the Moon . . .	`650  X3=(ZH-Z6+X0)*SC : Y3=SH*((Z6-X0)/ZH)*SC`
655		`655  X4=(ZH-Z7+X0)*SC : Y4=SH*((Z7-X0)/ZH)*SC`
660		`660  X3%=OX%+INT%(9*X3-8*X4) : X4%=OX%+INT%(9*X4-8*X3)`
665		`665  Y3%=OY%+INT%(9*Y3-8*Y4) : Y4%=OY%+INT%(9*Y4-8*Y3)`
670	. . . select red, and draw rotated path	`670  X=FNCOL(1) : X=FNLNE(X3%,Y3%,X4%,Y4%,RO,OX%,OY%)`
675	Skip next bit if lunar	`675  IF ET$="L" THEN GOTO 705`
680	R% is scaled radius of Moon	`680  R%=INT%(SC*RM)`
685		`685  X%=OX%+INT%((ZH-ZJ+X0)*SC)`
690		`690  Y%=OY%+INT%((SH*(ZJ-X0)/ZH)*SC)`
695	Select black and draw filled circle of Moon . . .	`695  X=FNCOL(0) : X=FNCIRCLE(X%,Y%,R%,1,RO,OX%,OY%)`
700	. . . then white and draw in Moon's outline	`700  X=FNCOL(7) : X=FNCIRCLE(X%,Y%,R%,0,RO,OX%,OY%)`
705	Lunar and solar; skip this bit if no rotation	`705  IF RW=0 THEN GOTO 720`
710	Select magenta, plot mode XOR . . .	`710  X2%=OX%+R1% : X=FNCOL(5) : X=FNPLOTM(2)`
715	. . . and draw radius of Sun or umbra in North direction	`715  X=FNLNE(OX%,OY%,X2%,OY%,TH,OX%,OY%) : X=FNPLOTM(0)`
720	Define window for the time-scale; select black . . .	`720  X=FNWIND(0,391,0,20) : X=FNCOL(0)`
725	. . . and clear window; reselect default settings	`725  X=FNWCOL : X=FNINIT`
730	Calculate . . .	`730  X=(ZH-Z1+X0)*SC : Y=SH*SC*(Z1-X0)/ZH`
735	. . . time-scale scale factor SC . . .	`735  A=FNM(RO) : B=COS(A) : SC=SC*B`
740		`740  FX=X*(1.0-B)+Y*SIN(A)`
745	. . . and time-scale offset XE%	`745  XE%=OX%+INT%(X+0.5)-INT%(FX+0.5)`
750	Select cyan, and draw in . . .	`750  X=FNCOL(6) : X=FNLNE(0,10,391,10,0.0,OX%,OY%)`
755	. . . two horizontal lines for time-scale	`755  X=FNLNE(0,13,391,13,0.0,OX%,OY%)`
760	Draw in 13 . . .	`760  FOR I=1 TO 13`
765		`765  X%=INT%((I-7+Z1-INT(Z1))*SC+0.5)+XE%`
770	. . . vertical pips to mark the hours	`770  X=FNLNE(X%,10,X%,17,0.0,OX%,OY%)`
775	Some are off the screen; select white	`775  NEXT I : X=FNCOL(7)`
780	Write 13 . . .	`780  FOR I=1 TO 13 : H=Z1+I-7`
785	. . . numbers to label the pips . . .	`785  IF H>24.0 THEN H=H-24.0`
790	. . . in the range 0–24 . . .	`790  IF H<0 THEN H=H+24.0`
795		`795  C$=STR$(INT%(H))`
800	. . . and label the scale as well	`800  IF I=7 THEN C$=C$+" UT"`
805		`805  X%=INT%((7-I+Z1-INT(Z1))*SC+0.5)+XE%`
810		`810  X=FNTEXT(X%,0,C$)`
815	Some numbers are off the screen	`815  NEXT I`
820	Define text box in the top left-hand corner . . .	`820  X=FNWIND(0,100,190,255) : X=FNCOL(0)`
825	. . . clear it, and select cyan	`825  X=FNWCOL : X=FNCOL(6)`
830	Draw a border around it . . .	`830  X=FNBOX(0,190,100,65) : X=FNCOL(7)`
835	. . . and draw in the text	`835  C$="LUNAR ECLIPSE"`
840		`840  IF ET$="S" THEN C$="SOLAR ECLIPSE"`
845		`845  X%=2 : Y%=245 : X=FNTEXT(X%,Y%,C$)`
850		`850  Y%=Y%-10 : X=FNTEXT(X%,Y%,MID$(DT$,3,12))`
855	Numbers in the D,M,S format via MINSEC	`855  SW(1)=1 : X=GL : GOSUB 1000`

```
860    C$="LO:"+MID$(OP$,3,13)
865    Y%=Y%-10 : X=FNTEXT(X%,Y%,C$)
870    X=FND(GP) : GOSUB 1000 : C$="LA:"+MID$(OP$,3,13)
875    Y%=Y%-10 : X=FNTEXT(X%,Y%,C$)
880    C$="HT: "+STR$(INT%(HT*6.37814E+06+0.5))+" M ASL"
885    Y%=Y%-10 : X=FNTEXT(X%,Y%,C$)
890    C$="UT: "+MID$(UE$,5,5)
895    Y%=Y%-10 : X=FNTEXT(X%,Y%,C$)
```

900	Print the picture?	900 Q$="Print (Y or N) .... "
905	YESNO gets the answer	905 PRINT : GOSUB 960
910	If not, normal return from DISPLAY	910 IF E=0 THEN PRINT : RETURN
915	Otherwise print it	915 X=FNPRNT
920	Another one? (the paper may have jammed)	920 Q$="Another print (Y or N) .... "
		925 GOSUB 960
		930 IF E=1 THEN GOTO 915
935	Return from DISPLAY	935 PRINT : RETURN

500 DISPLAY

Example

```
Lunar and solar eclipses
------------------------


Geographical longitude (W neg; D,M,S) .. ? -5,45,0
Geographical latitude (D,M,S) .........:. ? 50,0,0
Height above sea-level (m) ............. ? 0
Starting date (D,M,Y) .................. ? 11,8,1999
For how many months ................... ? 1

Solar eclipse certain

         Solar eclipse of  11  8  1999
         ---------------------------

UT of maximum eclipse .................     10 11
Eclipse begins at .....................      8 52
Eclipse ends at .......................     11 31
Eclipse magnitude .....................      1.004

Display the eclipse (Y or N) ........... ? Y
First contact (Y or N) ................. ? Y
Rotate ecliptic (Y or N) ............... ? Y

Azimuth (D,M,S; North = 0) ............. +108 34 40.67
Altitude (D,M,S) ......................  + 34 28  9.69
Ecliptic angle from horiz. (D,M,S CW) .. - 51 53 32.25
North angle from vertical (D,M,S CW) ... - 37 54 56.31

Print (Y or N) ........................ ? N

Display again (Y or N) ................? Y
First contact (Y or N) ................ ? N
Mid eclipse (Y or N) .................. ? Y
Rotate ecliptic (Y or N) .............. ? Y

Azimuth (D,M,S; North = 0) ............. +129  6 58.17
Altitude (D,M,S) ......................  + 45 39 31.49
Ecliptic angle from horiz. (D,M,S CW) .. - 38 52 47.88
North angle from vertical (D,M,S CW) ... - 31  2  5.07

Print (Y or N) ........................ ? N

Display again (Y or N) ................? N
Another location (Y or N) ............. ? N

No eclipse on  26  8  1999

Again (Y or N) ........................ ? N
```

7500 ELOSC

This routine calculates the instantaneous heliocentric ecliptic coordinates, radius vector and distance from the Earth, and the apparent geocentric coordinates and distance from the Earth (allowing for light-travel time), of any member of the solar system given its osculating elliptical orbital elements.

We saw in an earlier section how to calculate the position of any of the major planets using routine PLANS (4500) and its handling program. That routine itself called PELMENT (3800) to obtain the mean elliptical orbital elements for each of the planets, and then it applied lots of small corrections for the perturbing influences of other members of the Solar System. It was able to do this for each of the planets (except for Pluto) because dynamic theories of their motions have been worked out. It was therefore just a matter of plugging into the given formulae.

Most members of the Solar System, such as the periodic comets, minor planets, and Pluto, have not had dynamical theories worked out for their motions. Their orbital elements can be calculated on the basis of their observed orbits at any instant, but those elements change with time and will not be appropriate for a later instant. Elements which hold only over a short period near a given date are called *osculating elements*. If you use them to calculate positions within the period for which they are valid, you will get accurate results. The further you move outside that period, however, the more inaccurate the results become.

Osculating elements are given for the minor planets (and indeed for the major ones as well) in *The Astronomical Almanac*, correct for a specified instant. Having obtained them, you can use routine ELOSC and its handling program to calculate orbital positions. ELOSC is very similar to PLANS in many ways. However, it requires that you give it the elements (there is no equivalent of PELMENT) and it cannot allow for perturbations.

ELOSC accepts the elliptical osculating elements in the parameters EC (eccentricity), IN (inclination), OM (longitude of the ascending node), PE (longitude of the perihelion), AX (mean anomaly at the epoch), ND (mean daily motion), and PX (mean distance or semi-major axis, AU), at the epoch specified by DH (days), MH (months), and YH (years). All angles are in degrees. It calculates the instantaneous heliocentric ecliptic longitude, L0 (radians), ecliptic latitude, S0 (radians), radius vector, P0 (AU) and distance

7500 ELOSC

from the Earth, V0 (AU), at the instant specified by the date DY, MN, YR as usual. It also allows for light-travel time and returns the apparent geocentric ecliptic longitude, EP (radians), ecliptic longitude, BP (radians), and distance from the Earth, RH (AU).

The handling program HELOSC makes use of the input routine DEFAULT. This allows for default setting to be used instead of having to reenter all the numbers every time you pass through the program. For example, if you enter a value incorrectly, you do not need to start again with every number. Those values which are already correct can be handed back to the program by entering a comma in place of the digits (see the section on DEFAULT).

A further feature of the handling program HELOSC is that it can make allowance for the slowly changing equinoxes. You sometimes see the orbital elements referred to a particular equinox (e.g. 1950.0), but you wish to use them to find a position at a completely different time. HELOSC asks for the equinox date and corrects the given elements to your requirements by calling RELEM (7700) before calling ELOSC. You can then use, say, elements quoted for 1950.0 to find positions referred to the equinox of date.

Formulae
ELOSC uses the same formulae as PLANS (4500) to make its calculations, except that there are no corrections for perturbations.

Details of ELOSC
Called by GOSUB 7500.

For the date DY (days, including the fraction), MN (months), YR (years) calculates: L0, the heliocentric longitude, S0, the heliocentric latitude, P0, the radius vector of the planet, and V0, the distance from the Earth, not corrected for light time.

Also calculates EP, the geocentric ecliptic longitude, BP, the geocentric ecliptic latitude, and RH, the Earth–planet distance, corrected for light-travel time. All angles in *radians* and distances in *AU*.

Input parameters (besides the date) are: the eccentricity, EC, the mean anomaly at the epoch, AX, the mean daily motion in longitude, ND, the mean distance or semi-major axis, PX, the inclination, IN, the longitude of the ascending node, OM, the longitude of the perihelion, PE, and the epoch, DH (days), MN (months), YH (years). All input angles in *degrees*, and distances in *AU*.

Other routines called: JULDAY (1100), ANOMALY (3300), and SUN (3400).

```
                                    7497  REM
                                    7498  REM      Subroutine ELOSC
                                    7499  REM
```

7500	Initialise light-travel time LI=0	`7500  LI=0 : PI=3.1415926536 : TP=2.0*PI`
7505	Save date at which to calculate positions . . .	`7505  DO=DY : MO=MN : YO=YR : FL(1)=0`
7510	. . . and call JULDAY for DJ of epoch of elements . . .	`7510  DY=DH : MN=MH : YR=YH : GOSUB 1100`
7515	. . . saved in DD; now restore date and call SUN to find . . .	`7515  DD=DJ : FL(1)=0 : DY=DO : MN=MO : YR=YO : EZ=EC`
7520	. . . MS mean anomaly, LG Earth longitude, RE radius vector	`7520  GOSUB 3400 : MS=AM : RE=RR : LG=SR+PI : DZ=DJ-DD`
7525	First pass no light-travel time; second pass with light-travel time included	`7525  FOR K=1 TO 2`
7530	EC eccentricity; DZ days since epoch; call ANOMALY	`7530  EC=EZ : AM=FNM(AX+ND*DZ-LI*ND) : GOSUB 3300`
7535	PV radius vector	`7535  PV=PX*(1.0-EC*EC)/(1.0+EC*COS(AT))`
7540	AT is the true anomaly	`7540  LP=AT+FNM(PE) : LO=LP-FNM(OM)`
7545	Find trigonometric functions just once	`7545  SO=SIN(LO) : CO=COS(LO)`
		`7550  I1=FNM(IN) : SP=SO*SIN(I1) : Y=SO*COS(I1)`
		`7555  PS=FNS(SP) : PD=ATN(Y/CO)+FNM(OM)`
7560	Remove ambiguity of inverse tangent	`7560  IF CO<0 THEN PD=PD+PI`
		`7565  IF PD>TP THEN PD=PD-TP`
		`7570  CI=COS(PS) : RD=PV*CI : LL=PD-LG`
7575	RH is Earth–object distance	`7575  RH=RE*RE+PV*PV-2.0*RE*PV*CI*COS(LL)`
7580	LI is light-travel time	`7580  RH=SQR(RH) : LI=RH*5.775518E-3`
7585	Save true values if first pass	`7585  IF K=1 THEN LO=PD : SO=FNS(SP) : VO=RH : PO=PV`
7590	End of light-travel-time loop	`7590  NEXT K`
7595	Calculate trigonometric values once only	`7595  L1=SIN(LL) : L2=COS(LL)`
7600	Inner object, or outer?	`7600  IF PV<RE THEN GOTO 7610`
7605	Outer . . .	`7605  EP=ATN(RE*L1/(RD-RE*L2))+PD : GOTO 7615`
7610	. . . inner	`7610  EP=ATN(-1.0*RD*L1/(RE-RD*L2))+LG+PI`
7615	Make sure EP is in range 0–2π	`7615  IF EP<0 THEN EP=EP+TP : GOTO 7615`
		`7620  IF EP>TP THEN EP=EP-TP : GOTO 7620`
7625	Calculate geocentric ecliptic latitude	`7625  BP=ATN(RD*SP*SIN(EP-PD)/(CI*RE*L1))`
7630	Return from ELOSC	`7630  RETURN`

```
1    REM
2    REM        Handling program HELOSC
3    REM

5    DIM FL(20),ER(20),SW(20)
10   DEF FNI(W)=SGN(W)*INT(ABS(W))
15   DEF FNL(W)=FNI(W)+FNI((SGN(W)-1.0)/2.0)
20   DEF FNM(W)=1.745329252E-2*W
25   DEF FND(W)=5.729577951E1*W
30   DEF FNS(W)=ATN(W/(SQR(1-W*W)+1E-20))
35   DEF FNC(W)=1.570796327-FNS(W)
40   DEF FNQ$(W)=STR$(INT(W*1E4)/1E4)

45   PRINT : PRINT
50   PRINT "Osculating elliptical orbits"
55   PRINT "----------------------------"
60   PRINT : PRINT

65   QA$="Name of the object" : Q$=QA$ : X$=P$
70   N=0 : GOSUB 880 : P$=X$ : PRINT
75   Q$="Please input the orbital elements for "
80   PRINT Q$+P$+":" : PRINT

85   QB$="Eccentricity" : Q$=QB$ : X=EC
90   N=1 : GOSUB 880 : EC=X
95   QC$="Inclination (degrees)" : Q$=QC$ : X=IW
100  N=1 : GOSUB 880 : IW=X
105  QD$="Longitude of ascending node (deg.)"
110  Q$=QD$ : X=OW : N=1 : GOSUB 880 : OW=X

115  PRINT "Perihelion parameter:"

120  Q$="Specify argument (A) or longitude (L)"
125  UP$="A" : Q1$="Perihelion argument (degrees)"
130  UQ$="L" : Q2$="Perihelion longitude (degrees)"
135  X$=PP$ : GOSUB 625 : PP$=X$ : QE$=Q2$ : X=PW
140  IF E=1 THEN QE$=Q1$ : X=AW
145  Q$=QE$ : N=1 : GOSUB 880 : PP=X
150  IF E=0 THEN PW=X : AW=X-OW : GOTO 160
155  AW=X : PW=OW+X

160  QF$="Epoch (D,M,Y)" : Q$=QF$ : X=DH : Y=MH : Z=YH
165  N=3 : GOSUB 880 : DH=X : MH=Y : YH=Z

170  Q$="Specify mean anomaly (M) or longitude (L)"
175  UP$="M" : Q1$="Mean anomaly at epoch (degrees)"
180  UQ$="L" : Q2$="Longitude at epoch (degrees)"
185  X$=AQ$ : GOSUB 625 : AQ$=X$ : QG$=Q2$ : X=LW
190  IF E=1 THEN QG$=Q1$ : X=AX
195  Q$=QG$ : N=1 : GOSUB 880 : QW=X
200  IF E=0 THEN LW=X : AX=X-PW : GOTO 210
205  AX=X : LW=PW+X

210  Q$="Specify period (P) or daily motion (D)"

215  UP$="P" : Q1$="Orbital period (years)"
220  UQ$="D" : Q2$="Daily motion (degrees)"
225  X$=PD$ : GOSUB 625 : PD$=X$ : QH$=Q2$ : X=ND
230  IF E=1 THEN QH$=Q1$ : X=YP
235  Q$=QH$ : N=1 : GOSUB 880 : BT=X
240  IF E=0 THEN ND=X : GOTO 255
```

Notes column:

10 FNI is needed by JULDAY ...

15 ...and FNL by CALDAY

20 Converts degrees to radians ...

25 ...and radians to degrees

30 Returns inverse sine ...

35 ...and inverse cosine

40 Returns the value of W in string format to four decimal places

65 Get a name for neatness ...

75 ...and ask for its osculating elliptical elements ...

90 ...using DEFAULT for convenience

115 Can input perihelion position either as longitude ...

120 ...or as argument ...

135 ...call local routine to get number ...

150 ...and calculate the other choice

160 This is the epoch at which the longitudes have

165 ...the given values

170 Can input position at the epoch either as longitude ...

175 ...or as mean anomaly ...

185 ...call local routine to get number ...

200 ...and calculate the other choice

210 Can input speed of object either as mean daily motion ...

215 ...or as period of orbit ...

225 ...get number by calling local routine ...

245 ... and calculate daily motion if not given	245 `YP=X : ND=360.0/(YP*365.2422)`
	250 `PRINT Q2$; TAB(52); FNQ$(ND)`
	255 `QI$="Semi-major axis (mean distance; AU)"`
	260 `Q$=QI$ : X=PX : N=1 : GOSUB 880 : PX=X`
265 Display what you have entered ...	265 `PRINT : PRINT "Your values are:"`
	270 `PRINT "---------------" : PRINT`
	275 `PRINT QA$; TAB(52); P$`
	280 `PRINT QB$; TAB(52); FNQ$(EC)`
	285 `PRINT QC$; TAB(52); FNQ$(IW)`
	290 `PRINT QD$; TAB(52); FNQ$(OW)`
	295 `PRINT QE$; TAB(52); FNQ$(PP)`
	300 `PRINT QF$; TAB(51); INT(DH);INT(MH);INT(YH)`
	305 `PRINT QG$; TAB(52); FNQ$(QW)`
	310 `PRINT QH$; TAB(52); FNQ$(BT)`
	315 `PRINT QI$; TAB(52); FNQ$(PX)`
320 ... and ask for satisfaction ...	320 `Q$="Is this correct (Y or N) "`
325 ... with YESNO	325 `PRINT : GOSUB 960 : PRINT`
330 Mistakes are not too tedious because of DEFAULT	330 `IF E=0 THEN GOTO 65`
335 The date of the equinox to which the elements refer ...	335 `Q$="Equinox date of elements (D,M,Y)"`
	340 `X=DU : Y=MU : Z=YU`
	345 `N=3 : GOSUB 880 : DU=X : MU=Y : YU=Z`
350 ... and the date of the equinox to which the calculated ...	350 `Q$="Output for equinox (D,M,Y)"`
355 ... position should refer	355 `X=DB : Y=MB : Z=YB`
	360 `N=3 : GOSUB 880 : DB=X : MB=Y : YB=Z`
365 This is the date for which the position is required	365 `PRINT : Q$="Calculate position for date (D,M,Y)"`
	370 `X=DY : Y=MN : Z=YR : N=3 : GOSUB 880`
	375 `DY=X : MN=Y : YR=Z : PRINT`
380 Correct the elements to their values for the equinox ...	380 `X=FNM(IW) : Y=FNM(AW) : Z=FNM(OW)`
385 ... of date by calling RELEM	385 `GOSUB 7700 : IN=FND(P) : AR=FND(Q)`
	390 `OM=FND(R) : PE=AR+OM`
395 Make sure PE is in range 0–360 degrees	395 `IF PE>360.0 THEN PE=PE-360.0`
	400 `IF PE<0 THEN PE=PE+360.0`
405 DY carries time as fraction; call ELOSC	405 `UT=0 : GOSUB 7500`
410 Impossible date?	410 `IF ER(1)=1 THEN GOTO 575`
415 Display results ...	415 `PRINT "Heliocentric coordinates: "; P$`
	420 `PRINT "----------- -----------" : PRINT`
425 ... in D,M,S format via MINSEC	425 `SW(1)=1 : NC=9 : X=FND(L0) : GOSUB 1000`
	430 `Q$="Ecliptic longitude (D,M,S) .......... "`
	435 `PRINT Q$+OP$ : X=FND(S0) : GOSUB 1000`
	440 `Q$="Ecliptic latitude (D,M,S) ........... "`
	445 `PRINT Q$+OP$`
	450 `Q$="Radius vector (AU) .................. "`
455 Radius vector to six decimal places	455 `PRINT Q$; " "+STR$(INT(1E6*P0+0.5)/1E6)`
	460 `PRINT : PRINT "Geocentric coordinates: "; P$`
	465 `PRINT "--------- -----------" : PRINT`
	470 `Q$="Ecliptic longitude (D,M,S) .......... "`
	475 `X=FND(EP) : GOSUB 1000 : PRINT Q$+OP$`
	480 `Q$="Ecliptic latitude (D,M,S) ........... "`
	485 `X=FND(BP) : GOSUB 1000 : PRINT Q$+OP$`
	490 `Q$="Distance from Earth (AU) ............ "`
495 Distance from Earth to six decimal places	495 `PRINT Q$; " "+STR$(INT(1E6*V0+0.5)/1E6)`
500 Correct geocentric ecliptic coordinates for aberration	500 `A=LG+PI-EP : B=COS(A) : C=SIN(A)`
	505 `EP=EP-(9.9387E-5*B/COS(BP))`
	510 `BP=BP-(9.9387E-5*C*SIN(BP))`

201

7500 ELOSC

515 Set up date of equinox for output

515	Set up date of equinox for output
520	Turn off nutation by setting FL(6)=1
525	Call EQECL for right ascension and declination, then call PRCESS1
530	Restore date
560	Calculate the solar elongation, and display...
570	...to one decimal place
575	Another go?
580	YESNO tells us the answer
595	Reset the flags: 1: JULDAY, 7: EQECL, 10: PRCESS1
625	Local input routine; get character via DEFAULT
630	Respond to answer in upper or lower case
640	Ask again if answer is not recognised

```
515   DA=DY : MA=MN : YA=YR : DY=DB : MN=MB : YR=YB
520   X=EP : Y=BP : SW(3)=-1 : FL(6)=1
525   GOSUB 2000 : X=P : Y=Q : GOSUB 2500
530   DY=DA : MN=MA : YR=YA
535   P=FND(P) : Q=FND(Q) : X=P/15.0 : GOSUB 1000
540   Q$="Astrometric right ascension (H,M,S) .. "
545   PRINT Q$+"   "+MID$(OP$,4,12)
550   Q$="Astrometric declination (D,M,S) ...... "
555   X=Q : GOSUB 1000 : PRINT Q$+OP$
560   D=-COS(BP)*COS(EP-LG) : E=FND(FNC(D)) : PRINT
565   Q$="Solar elongation (degrees) .......... "
570   PRINT Q$; "   "+STR$(INT(10.0*E+0.5)/10.0)
575   Q$="Again (Y or N) ..................... "
580   PRINT : GOSUB 960
585   IF E=0 THEN STOP
590   Q$="New object (Y or N) ................. "
595   GOSUB 960 : FL(1)=0 : FL(7)=0 : FL(10)=0
600   IF E=1 THEN PRINT : GOTO 65
605   Q$="New equinoxes (Y or N) .............. "
610   GOSUB 960
615   IF E=1 THEN PRINT : GOTO 335
620   GOTO 365
625   N=0 : GOSUB 880
630   IF X$=UP$ OR X$=CHR$(ASC(UP$)+32) THEN E=1 : RETURN
635   IF X$=UQ$ OR X$=CHR$(ASC(UQ$)+32) THEN E=0 : RETURN
640   PRINT "What ? " : GOTO 625

INCLUDE DEFAULT, YESNO, MINSEC, JULDAY, OBLIQ, NUTAT,
        EQECL, PRCESS1, ANOMALY, SUN, ELOSC, RELEM
```

Example

```
Osculating elliptical orbits
----------------------------


Name of the object []                       ? CERES

Please input the orbital elements for CERES:

Eccentricity [0]                            ? 0.0784
Inclination (degrees) [0]                   ? 10.606
Longitude of ascending node (deg.) [0]      ? 80.718
Perihelion parameter:
Specify argument (A) or longitude (L) []    ? A
Perihelion argument (degrees) [0]           ? 72.890
Epoch (D,M,Y) [0,0,0]                        ? 27,10,1984
Specify mean anomaly (M) or longitude (L) [] ? M
Mean anomaly at epoch (degrees) [0]         ? 260.117
Specify period (P) or daily motion (D) []   ? D
Daily motion (degrees) [0]                  ? 0.21419
Semi-major axis (mean distance; AU) [0]     ? 2.7666
```

```
Your values are:
----------------

Name of the object                             CERES
Eccentricity                                   0.0784
Inclination (degrees)                          10.606
Longitude of ascending node (deg.)             80.718
Perihelion argument (degrees)                  72.89
Epoch (D,M,Y)                                  27  10  1984
Mean anomaly at epoch (degrees)                260.117
Daily motion (degrees)                         0.2141
Semi-major axis (mean distance; AU)            2.7666

Is this correct (Y or N) ? Y

Equinox date of elements (D,M,Y) [0,0,0]     ? 1.5,1,2000
Output for equinox (D,M,Y) [0,0,0]           ? 1.5,1,2000

Calculate position for date (D,M,Y) [0,0,0]   ? 21,7,1984

Heliocentric coordinates: CERES
------------ ------------

Ecliptic longitude (D,M,S) ..........    + 25 39 37.95
Ecliptic latitude (D,M,S) ............   -  8 42 17.77
Radius vector (AU) ...................   2.88968

Geocentric coordinates: CERES
---------- ------------

Ecliptic longitude (D,M,S) ..........    + 45 31 56.66
Ecliptic latitude (D,M,S) ............   -  8 20  5.74
Distance from Earth (AU) ............    3.017176
Astrometric right ascension (H,M,S) ..     3  2 51.13
Astrometric declination (D,M,S) ......   +  8 33 51.95

Solar elongation (degrees) ..........    73.1

Again (Y or N) ...................... ? Y
New object (Y or N) ................. ? Y

Name of the object [CERES]                   ? PALLAS

Please input the orbital elements for PALLAS:

Eccentricity [0.0784]                        ? 0.2335
Inclination (degrees) [10.606]               ? 34.794
Longitude of ascending node (deg.) [80.718]  ? 173.347
Perihelion parameter:
Specify argument (A) or longitude (L) [A]    ? ,
Perihelion argument (degrees) [72.89]        ? 309.930
Epoch (D,M,Y) [27,10,1984]                   ? ,,,
Specify mean anomaly (M) or longitude (L) [M] ? ,
Mean anomaly at epoch (degrees) [260.117]    ? 249.109
Specify period (P) or daily motion (D) [D]   ? ,
Daily motion (degrees) [0.2141]              ? 0.21361
Semi-major axis (mean distance; AU) [2.7666] ? 2.7715

Your values are:
----------------

Name of the object                             PALLAS
Eccentricity                                   0.2335
Inclination (degrees)                          34.794
Longitude of ascending node (deg.)             173.347
```

7500 ELOSC

```
Perihelion argument (degrees)                    309.93
Epoch (D,M,Y)                                    27  10  1984
Mean anomaly at epoch (degrees)                  249.109
Daily motion (degrees)                           0.2136
Semi-major axis (mean distance; AU)              2.7715

Is this correct (Y or N) ? Y

Equinox date of elements (D,M,Y) [1.5,1,2000]    ? ,,,
Output for equinox (D,M,Y) [1.5,1,2000]          ? ,,,

Calculate position for date (D,M,Y) [21,7,1984]  ? ,,,

Heliocentric coordinates: PALLAS
------------ ------------

Ecliptic longitude (D,M,S) ...........    +337 53 44.95
Ecliptic latitude (D,M,S) ............    + 10 20 56.48
Radius vector (AU) ...................    3.269845

Geocentric coordinates: PALLAS
---------- ------------

Ecliptic longitude (D,M,S) ...........    +352 46  1.38
Ecliptic latitude (D,M,S) ............    + 13  8 22.83
Distance from Earth (AU) .............    2.584315
Astrometric right ascension (H,M,S) ..     23 13 20.75
Astrometric declination (D,M,S) ......    +  9 16 46.64

Solar elongation (degrees) ...........    124.6

Again (Y or N) ...................... ? N
```

The Astronomical Almanac gives the J 2000.0 astronometric positions of Ceres and Pallas at 1984 July 21.0 as follows:

	Right ascension	Declination	True distance
Ceres	03 02 51.7	08 33 54	3.017
Pallas	23 13 20.9	09 16 43	2.584

7700 RELEM

This routine converts ecliptic orbital elements to one equinox to their corresponding values referred to another.

It is sometimes necessary to convert the orbital elements of a planet, a minor planet, or a comet from one equinox to another. There are three elements affected by the changing equinox, namely the inclination, the argument of the perihelion, and the longitude of the ascending node. The other elements are not affected.

RELEM carries out the reduction of these elements from epoch 1, specified by the variables DU, MU, YU to epoch 2, specified by DY, MN, YR. The elements are input via the variables X, Y, and Z corresponding to the inclination, argument of the perihelion, and longitude of the ascending node respectively at epoch 1, and are returned via P, Q, and R correct for epoch 2. On returning from the routine, the current value of DJ (Julian days since 1900 January 0.5) corresponds to epoch 2, and FL(1) is set to 1.

Formulae

$$\text{SIN(P)} \times \text{SIN(R} - \text{TH)} = \text{SIN(X)} \times \text{SIN(Z} - \text{TA)}$$
$$\text{SIN(P)} \times \text{COS(R} - \text{TH)} = \text{COS(ET)} \times \text{SIN(X)} \times \text{COS(Z} - \text{TA)}$$
$$- \text{SIN(ET)} \times \text{COS(X)}$$
$$\text{SIN(P)} \times \text{SIN(DQ)} = \text{SIN(ET)} \times \text{SIN(Z} - \text{TA)}$$
$$\text{SIN(P)} \times \text{COS(DQ)} = \text{SIN(X)} \times \text{COS(ET)} - \text{COS(X)} \times \text{SIN(ET)}$$
$$\times \text{COS(Z} - \text{TA)}$$

$$Q = Y + DQ$$
$$TT = (DJ - D1)/365250$$
$$T0 = D1/365250$$

DJ = Julian days since 1900 January 0.5 to epoch 2
D1 = Julian days since 1900 January 0.5 to epoch 1

ET, TH, and TA are determined from series of terms involving TT and T0 (see BASIC listing).

7700 RELEM

Details of RELEM

Called by GOSUB 7700.

Calculates P, the inclination, Q, the argument of the perihelion, and R, the longitude of the ascending node, correct at epoch 2, DY (days), MN (months), YR (years), given the corresponding elements X, Y, and Z, correct at epoch 1, DU, MU, YU. All angles in *radians*.

Other routine called: JULDAY (1100).

	7697 REM
	7698 REM Subroutine RELEM
	7699 REM
7700 Save epoch 2; set up epoch 1	7700 DO=DY : MO=MN : YO=YR : DY=DU : MN=MU : YR=YU
7705 Call JULDAY and find T0 for epoch 1; save error flag	7705 FL(1)=0 : GOSUB 1100 : TO=DJ/365250.0 : E1=ER(1)
7710 Set up epoch 2, and call JULDAY	7710 DY=DO : MN=MO : YR=YO : FL(1)=0 : GOSUB 1100
7715 Find T1 for epoch 2, and various . . .	7715 T1=DJ/365250.0 : PI=3.141592654 : TT=T1-T0
7720 . . . other bits; combine previous error flag in current one	7720 T2=TT*TT : T3=T0*T0 : T4=TT*T2 : ER(1)=ER(1)+E1
7725 Calculate ET . . .	7725 ET=(471.07-6.75*T0+0.57*T3)*TT+(0.57*T0-3.37)*T2
	7730 ET=ET+0.05*T4 : ET=FNM(ET/3600.0)
7735 . . . TA . . .	7735 TA=32869.0*T0+56.0*T3-(8694.0+55.0*T0)*TT+3.0*T2
	7740 TA=FNM((TA/3600.0)+173.950833)
7745 . . . and TH in radians	7745 TH=(50256.41+222.29*T0+0.26*T3)*TT
	7750 TH=TH+(111.15+0.26*T0)*T2+0.1*T4
	7755 TH=TA+FNM(TH/3600)
7760 Calculate these trigonometric quantities just once	7760 CI=COS(X) : SI=SIN(X) : OT=Z-TA : SO=SIN(OT)
	7765 CO=COS(OT) : SJ=SIN(ET) : CJ=COS(ET) : A=SI*SO
	7770 B=CJ*SI*CO-SJ*CI : O1=ATN(A/B)
7775 Remove ambiguity of inverse tangent . . .	7775 IF B<0 THEN O1=O1+PI
	7780 B=SI*CJ-CI*SJ*CO : A=-SJ*SO : DC=ATN(A/B)
7785 . . . here as well	7785 IF B<0 THEN DC=DC+PI
7790 Call local routine to reduce to primary interval	7790 W=Y+DC : TP=2.0*PI : GOSUB 7815 : Q=W
	7795 W=O1+TH : GOSUB 7815 : R=W
7800 Use best formula for P depending on the value of X	7800 IF X<0.175 THEN P=FNS(A/SIN(DC)) : RETURN
	7805 CI=(CI*CJ)+(SI*SJ*CO) : P=1.570796327-FNS(CI)
7810 Return from RELEM	7810 RETURN
7815 Local routine to reduce W . . .	7815 IF W<0 THEN W=W+TP : GOTO 7815
7820 . . . to the range $0-2\pi$	7820 IF W>TP THEN W=W-TP : GOTO 7820
	7825 RETURN

```
1      REM
2      REM        Handling program HRELEM
3      REM

5      DIM FL(20),ER(20),SW(20)
10     DEF FNI(W)=SGN(W)*INT(ABS(W))

15     DEF FNM(W)=1.745329252E-2*W
20     DEF FND(W)=5.729577951E1*W
25     DEF FNS(W)=ATN(W/(SQR(1-W*W)+1E-20))

30     PRINT : PRINT : PRINT
35     PRINT "Reduction of ecliptic elements"
40     PRINT "---------------------------"
45     PRINT

50     Q$="Elements specified at epoch (D,M,Y) ...... "
55     PRINT : PRINT Q$; : INPUT DU,MU,YU
60     Q$="Elements to be reduced to epoch (D,M,Y) .. "
65     PRINT Q$; : INPUT DY,MN,YR

70     Q1$="Inclination (degrees) ................... "
75     PRINT : PRINT Q1$; : INPUT X
80     Q2$="Argument of perihelion (degrees) ........ "
85     PRINT Q2$; : INPUT Y
90     Q3$="Longitude of ascending node (degrees) .... "
95     PRINT Q3$; : INPUT Z

100    X=FNM(X) : Y=FNM(Y) : Z=FNM(Z) : GOSUB 7700 : PRINT

105    IF ER(1)<>0 THEN GOTO 135

110    PRINT "Reduced coordinates are:"
115    PRINT "----------------------" : PRINT
120    P=FND(P) : PRINT Q1$+"  "+STR$(INT(P*1E6)/1E6)
125    Q=FND(Q) : PRINT Q2$+"  "+STR$(INT(Q*1E6)/1E6)
130    R=FND(R) : PRINT Q3$+"   "+STR$(INT(R*1E6)/1E6)

135    Q$="Again (Y or N) ......................... "
140    PRINT : GOSUB 960
145    IF E=0 THEN STOP
150    Q$="New epochs (Y or N) ................... "
155    GOSUB 960
160    IF E=1 THEN GOTO 50
165    GOTO 70

INCLUDE YESNO, JULDAY, RELEM
```

Notes column:

10   FNI returns the truncated-integer function
15   FNM converts degrees to radians . . .
20   . . . and FND converts radians to degrees
25   FNS returns the inverse sine

50   Get the two epochs . . .
55   . . . epoch 1 . . .
65   . . . and epoch 2

70   These are the elements specified at epoch 1

100   Convert elements to radians, and call RELEM
105   Impossible dates?

110   Display the results . . .

120   . . . in decimal degrees to six decimal places

135   Another reduction?
140   Call YESNO to find out

150   Either new epochs and new elements . . .

165   . . . or just new elements

Example

```
Reduction of ecliptic elements
------------------------------

Elements specified at epoch (D,M,Y) ...... ? 1.5,1,2000
Elements to be reduced to epoch (D,M,Y) .. ? 27,10,1984

Inclination (degrees) .................... ? 34.794
Argument of perihelion (degrees) ......... ? 309.930
Longitude of ascending node (degrees) .... ? 173.347

Reduced coordinates are:
------------------------

Inclination (degrees) ....................    34.795982
Argument of perihelion (degrees) .........   309.929905
Longitude of ascending node (degrees) ....   173.135076

Again (Y or N) ........................... ? Y
New epochs (Y or N) ...................... ? N

Inclination (degrees) .................... ? 10.606
Argument of perihelion (degrees) ......... ? 72.890
Longitude of ascending node (degrees) .... ? 80.718

Reduced coordinates are:
------------------------

Inclination (degrees) ....................    10.605855
Argument of perihelion (degrees) .........    72.87925
Longitude of ascending node (degrees) ....    80.516563

Again (Y or N) ........................... ? N
```

7900 PCOMET

This routine calculates the instantaneous heliocentric ecliptic coordinates, radius vector and distance from the Earth, and the apparent geocentric ecliptic coordinates and distance from the Earth (allowing for light-travel time), of a comet given its parabolic orbital elements.

Comets are members of our Solar System, usually with highly elongated orbits, which become visible near the Sun. They have bright heads and diffuse tails of variable length which always point away from the Sun. Comets can be divided into two categories: those which are gravitationally bound to the Sun like the planets and travel in elliptical orbits, the *periodic* comets, and those which do not seem to be bound to the Sun but appear once and then shoot off into space never to be seen again, the *parabolic* comets. Routine PCOMET deals with the latter sort; you can use routine ELOSC (7500) for periodic comets.

The approximate position of a *parabolic* comet can be found from its parabolic orbital elements. As the name suggests, the orbit is a parabola and needs one less element to define it than does an ellipse. Subroutine PCOMET (7900) performs this task, taking the five parabolic orbital elements and calculating the ecliptic coordinates of the comet for a given instant. This is input in the usual way via DY, MN, YR, returning with the ecliptic longitude, EP, the latitude, BP, and the distance of the comet from the Earth, RH. These quantities have all been corrected for light-travel time. The routine also returns the instantaneous values (i.e. not corrected for light-travel time) of the heliocentric ecliptic longitude, L0, latitude, S0, radius vector, P0, and distance from Earth, V0.

PCOMET expects the orbital elements to be supplied by the calling program as follows: inclination, IN, longitude of the ascending node, OM, argument of the perihelion, AR, the epoch of the perihelion, DI, MI, YI, and the perihelion distance, QP. Note that the longitude of the perihelion is equal to the argument of the perihelion plus the longitude of the ascending node.

The parabolic elements are usually quoted referred to some standard epoch, often 1950.0. PCOMET, on the other hand, expects the elements for the equinox of date. You can convert the elements from one epoch to another by means of subroutine RELEM (7700). The handling program, HPCOMET, listed below does this automatically. You input the parabolic elements referred to a given epoch, and it calculates the comet's position at the given date but referred to any epoch you specify. Thus, you can input orbital elements referred to the equinox of 1950.0, and calculate the position at, say 12th May 1994, referred to the equinox of J2000.0.

HPCOMET makes use of routine DEFAULT to ease the process of entering lots of data when running the program. When you enter a number, say OM = 123.456, that number becomes the default value for OM which the computer will use the next time round without your having to reenter it. Thus if you make a mistake when entering a long list of numbers, you do not need to start again, but can reuse the correct values, changing only those which are wrong. See DEFAULT (880).

Formulae

$W = ((DJ - LI - DQ) \times 3.649116E{-}2)/(QP^{1.5})$

DJ = required instant in Julian days since 1900 January 0.5

DQ = Julian days since 1900 January 0.5 of perihelion epoch

LI = light-travel time (days)

$0 = S^3 + (3 \times S) - W$ (parabolic equation)

$S = TAN(NU/2)$

$R = QP \times (1 + S^2)$ = radius vector of comet

NU = true anomaly.

Then the calculation continues as for an elliptical orbit. The parabolic equation is solved by an iterative method to the accuracy specified in line 7935 (1E–6 here).

Details of PCOMET

Called by GOSUB 7900.

For the date DY (days, including the fraction), MN (months), YR (years) calculates: L0, the heliocentric longitude, S0, the heliocentric latitude, P0, the radius vector of the planet, and V0, the distance from the Earth, not corrected for light-travel time.

Also calculates EP, the geocentric ecliptic longitude, BP, the geocentric ecliptic latitude, and RH, the Earth–comet distance, corrected for light-travel time. All angles in *radians* and distances in *AU*.

Input parameters (besides the date) are: the inclination, IN, the longitude of the ascending node, OM, the argument of the perihelion, AR, the epoch of the perihelion, DI (days, including the fraction), MI (months), YI (years), and the perihelion distance, QP. All input angles in *degrees*, and distances in *AU*.

Other routines called: JULDAY (1100), and SUN (3400).

```
7897 REM
7898 REM        Subroutine PCOMET
7899 REM
```

Line	Note	Code
7900	Save input date; set up for epoch of perihelion	`7900 DO=DY : MO=MN : YO=YR : DY=DI : MN=MI : YR=YI`
7905	Call JULDAY, and set DQ equal to the result	`7905 FL(1)=0 : GOSUB 1100 : DQ=DJ : DY=DO : MN=MO`
7910	Save error flag; set up input date, and call SUN	`7910 YR=YO : E1=ER(1) : FL(1)=0 : GOSUB 3400`
7915	LG is longitude of Earth; RE is Sun–Earth distance	`7915 ER(1)=ER(1)+E1 : LG=SR+PI : RE=RR`
7920	LI is light-travel time; make allowance for it on second pass	`7920 LI=0 : FOR K=1 TO 2`
7925	Loop to find true anomaly, NU . . .	`7925 W=(DJ-LI-DQ)*3.649116E-2/(QP*SQR(QP)) : S=W/3.0`

7935	. . . until we have the required accuracy (1E–6)	`7930 S2=S*S : D=(S2+3.0)*S-W`
		`7935 IF ABS(D)<1E-6 THEN GOTO 7945`
		`7940 S=((2.0*S*S2)+W)/(3.0*(S2+1.0)) : GOTO 7930`

7945	Get NU from S; R is radius vector of comet	`7945 NU=2.0*ATN(S) : R=QP*(1.0+S2) : L=NU+FNM(AR)`

7950	Find trigonometric values just once	`7950 S1=SIN(L) : C1=COS(L) : I1=FNM(IN)`
		`7955 S2=S1*SIN(I1) : PS=FNS(S2) : Y=S1*COS(I1)`
		`7960 LC=ATN(Y/C1)+FNM(OM) : C2=COS(PS)`

7965	Correct for ambiguity of inverse tangent	`7965 IF C1<0 THEN LC=LC+PI` `7970 IF LC>TP THEN LC=LC-TP : GOTO 7970`
7980	RH is distance of comet from the Earth	`7975 RD=R*C2 : LL=LC-LG : C3=COS(LL) : S3=SIN(LL)` `7980 RH=SQR((RE*RE)+(R*R)-(2*RE*RD*C3))`
7985	Save the instantaneous values	`7985 IF K=1 THEN LO=LC : SO=PS : VO=RH : PO=R`
7990	Calculate the light-travel time, and repeat	`7990 LI=RH*5.775518E-3 : NEXT K`
7995	Inner comet or outer?	`7995 IF RD<RE THEN GOTO 8005`
8000	Outer . . .	`8000 EP=ATN((RE*S3)/(RD-(RE*C3)))+LC : GOTO 8010`
8005	. . . inner	`8005 EP=ATN((-RD*S3)/(RE-(RD*C3)))+LG+PI`
8010	Make sure EP is in range 0–2π	`8010 IF EP<0 THEN EP=EP+TP : GOTO 8010` `8015 IF EP>TP THEN EP=EP-TP : GOTO 8015`
8025	Return from PCOMET	`8020 TB=(RD*S2*SIN(EP-LC))/(C2*RE*S3)` `8025 BP=ATN(TB) : RETURN`

```
1    REM
2    REM          Handling program HPCOMET
3    REM

5    DIM FL(20),ER(20),SW(20)
10   DEF FNI(W)=SGN(W)*INT(ABS(W))

15   DEF FNL(W)=FNI(W)+FNI((SGN(W)-1.0)/2.0)

20   DEF FNM(W)=1.745329252E-2*W
25   DEF FND(W)=5.729577951E1*W
30   DEF FNS(W)=ATN(W/(SQR(1-W*W)+1E-20))
35   DEF FNC(W)=1.570796327-FNS(W)
40   DEF FNQ$(W)=STR$(INT(W*1E4)/1E4)

45   PRINT : PRINT
50   PRINT "Parabolic orbits"
55   PRINT "----------------" : PRINT : PRINT

60   QA$="Name of the object" : Q$=QA$ : X$=P$
65   N=0 : GOSUB 880 : P$=X$ : PRINT
70   Q$="Please input the orbital elements for "
75   PRINT Q$+P$+":" : PRINT

80   QC$="Inclination (degrees)" : Q$=QC$ : X=IW
85   N=1 : GOSUB 880 : IW=X
90   QD$="Longitude of ascending node (deg.)"
95   Q$=QD$ : X=OW : N=1 : GOSUB 880 : OW=X
100  QE$="Argument of perihelion (degrees)"
105  Q$=QE$ : N=1 : X=AW : GOSUB 880 : AW=X
110  QF$="Epoch of perihelion (D,M,Y)" : Q$=QF$
115  X=DI : Y=MI : Z=YI : N=3 : GOSUB 880
120  DI=X : MI=Y : YI=Z
125  QG$="Perihelion distance (AU)"
130  Q$=QG$ : X=QP : N=1 : GOSUB 880 : QP=X

135  PRINT : PRINT "Your values are:"
140  PRINT "----------------" : PRINT
145  PRINT QA$; TAB(52); P$
150  PRINT QC$; TAB(52); FNQ$(IW)
155  PRINT QD$; TAB(52); FNQ$(OW)
160  PRINT QE$; TAB(52); FNQ$(AW)
165  PRINT QF$; TAB(52); FNQ$(DI);INT(MI);INT(YI)
170  PRINT QG$; TAB(52); FNQ$(QP)

175  Q$="Is this correct (Y or N) "
180  PRINT : GOSUB 960 : PRINT
185  IF E=0 THEN GOTO 60

190  Q$="Equinox date of elements (D,M,Y)"

195  X=DU : Y=MU : Z=YU
200  N=3 : GOSUB 880 : DU=X : MU=Y : YU=Z
205  Q$="Output for equinox (D,M,Y)"

210  X=DB : Y=MB : Z=YB
215  N=3 : GOSUB 880 : DB=X : MB=Y : YB=Z

220  PRINT : Q$="Calculate position for date (D,M,Y)"
225  X=DY : Y=MN : Z=YR : N=3 : GOSUB 880
230  DY=X : MN=Y : YR=Z : PRINT

235  X=FNM(IW) : Y=FNM(AW) : Z=FNM(OW)
```

10	FNI returns the truncated-integer function . . .
15	. . . and FNL returns the least-integer function
20	FNM converts degrees to radians . . .
25	. . . and FND converts radians to degrees
30	FNS returns the inverse sine . . .
35	. . . and FNC returns the inverse cosine
40	FNQ returns the value of W in string format to four decimal places
60	Get a name for clarity . . .
65	. . . using DEFAULT for all input
80	Get the parabolic elements . . .
85	. . . using DEFAULT
135	Report the values . . .
150	. . . to four decimal places (FNQ$)
175	Any mistakes?
180	Ask with YESNO
185	Correct mistakes if any
190	The elements were referred to the equinox . . .
195	. . . of this date . . .
205	. . . and the position is to be calculated . . .
210	. . . using the equinox of this date
220	We want to know where the comet is . . .
225	. . . on this date
235	Convert elements to radians . . .

240	. . . and call RELEM to correct for different equinoxes	240	`GOSUB 7700 : IN=FND(P) : AR=FND(Q) : OM=FND(R)`
245	Set UT=0 (time carried as fraction of day by DY) . . .	245	`UT=0 : GOSUB 7900`
250	. . . and call PCOMET; impossible date?	250	`IF ER(1)=1 THEN GOTO 415`
255	Display the results . . .	255	`PRINT "Heliocentric coordinates: "; P$`
		260	`PRINT "----------- -----------" : PRINT`
265	. . . in D,M,S format using MINSEC	265	`SW(1)=1 : NC=9 : X=FND(L0) : GOSUB 1000`
		270	`Q$="Ecliptic longitude (D,M,S) .......... "`
		275	`PRINT Q$+OP$ : X=FND(S0) : GOSUB 1000`
		280	`Q$="Ecliptic latitude (D,M,S) ........... "`
		285	`PRINT Q$+OP$`
		290	`Q$="Radius vector (AU) ................... "`
295	. . . to six decimal places	295	`PRINT Q$; " "+STR$(INT(1E6*P0+0.5)/1E6)`
		300	`PRINT : PRINT "Geocentric coordinates: "; P$`
		305	`PRINT "---------- -----------" : PRINT`
		310	`Q$="Ecliptic longitude (D,M,S) .......... "`
		315	`X=FND(EP) : GOSUB 1000 : PRINT Q$+OP$`
		320	`Q$="Ecliptic latitude (D,M,S) ........... "`
		325	`X=FND(BP) : GOSUB 1000 : PRINT Q$+OP$`
		330	`Q$="Distance from Earth (AU) ............ "`
335	. . . to six decimal places	335	`PRINT Q$; " "+STR$(INT(1E6*V0+0.5)/1E6)`
340	Correct for aberration	340	`A=LG+PI-EP : B=COS(A) : C=SIN(A)`
		345	`EP=EP-(9.9387E-5*B/COS(BP))`
		350	`BP=BP-(9.9387E-5*C*SIN(BP))`
355	Set up the epochs . . .	355	`DA=DY : MA=MN : YA=YR : DY=DB : MN=MB : YR=YB`
360	. . . turn off nutation (FL(6)=1) . . .	360	`X=EP : Y=BP : SW(3)=-1 : FL(6)=1`
365	. . . call EQECL for right ascension and declination, and PRCESS1	365	`GOSUB 2000 : X=P : Y=Q : GOSUB 2500`
370	Restore date	370	`DY=DA : MN=MA : YR=YA`
375	Display results in H,M,S format with MINSEC	375	`P=FND(P) : Q=FND(Q) : X=P/15.0 : GOSUB 1000`
		380	`Q$="Astrometric right ascension (H,M,S) .. "`
		385	`PRINT Q$+"    "+MID$(OP$,4,12)`
		390	`Q$="Astrometric declination (D,M,S) ...... "`
		395	`X=Q : GOSUB 1000 : PRINT Q$+OP$`
400	Calculate the elongation . . .	400	`D=-COS(BP)*COS(EP-LG) : E=FND(FNC(D)) : PRINT`
		405	`Q$="Solar elongation (degrees) .......... "`
410	. . . and display it to one decimal place	410	`PRINT Q$; "   "+STR$(INT(10.0*E+0.5)/10.0)`
415	Another go?	415	`Q$="Again (Y or N) ...................... "`
420	Get answer with YESNO	420	`PRINT : GOSUB 960`
		425	`IF E=0 THEN STOP`
		430	`Q$="New object (Y or N) ................. "`
435	Reset flags: 1: JULDAY, 7: EQECL, 10: PRCESS1	435	`GOSUB 960 : FL(1)=0 : FL(7)=0 : FL(10)=0`
		440	`IF E=1 THEN PRINT : GOTO 60`
		445	`Q$="New equinoxes (Y or N) .............. "`
		450	`GOSUB 960`
		455	`IF E=1 THEN PRINT : GOTO 190`
		460	`GOTO 220`

```
INCLUDE DEFAULT, YESNO, MINSEC, JULDAY, OBLIQ, NUTAT,
        EQECL, PRCESS1, ANOMALY, SUN, RELEM, PCOMET
```

Example

```
Parabolic orbits
----------------


Name of the object []                          ? Comet Kohler (1977m)

Please input the orbital elements for Comet Kohler (1977m):

Inclination (degrees) [0]                      ? 48.7196
Longitude of ascending node (deg.) [0]         ? 181.8175
Argument of perihelion (degrees) [0]           ? 163.4799
Epoch of perihelion (D,M,Y) [0,0,0]            ? 10.5659,11,1977
Perihelion distance (AU) [0]                   ? 0.990662

Your values are:
----------------


Name of the object                             Comet Kohler (1977m)
Inclination (degrees)                          48.7196
Longitude of ascending node (deg.)             181.8175
Argument of perihelion (degrees)               163.4799
Epoch of perihelion (D,M,Y)                    10.5659 11  1977
Perihelion distance (AU)                       0.9906

Is this correct (Y or N) ? Y

Equinox date of elements (D,M,Y) [0,0,0]       ? 0.9,1,1950
Output for equinox (D,M,Y) [0,0,0]             ? 20,5,1977

Calculate position for date (D,M,Y) [0,0,0]    ? 20,5,1977

Heliocentric coordinates: Comet Kohler (1977m)
------------ ------------


Ecliptic longitude (D,M,S) ...........  +227 24 43.62
Ecliptic latitude (D,M,S) ............  + 38 56 56.23
Radius vector (AU) ...................  2.780344

Geocentric coordinates: Comet Kohler (1977m)
---------- ------------


Ecliptic longitude (D,M,S) ...........  +217 39 39.98
Ecliptic latitude (D,M,S) ............  + 55 48  1.79
Distance from Earth (AU) .............  2.113133
Astrometric right ascension (H,M,S) ..    15 41 28.76
Astrometric declination (D,M,S) ......  + 38 28 34.29

Solar elongation (degrees) ...........  121.6

Again (Y or N) ......................  ? Y
New object (Y or N) ..................  ? N
New equinoxes (Y or N) ..............  ? Y

Equinox date of elements (D,M,Y) [0.9,1,1950]  ? ,,,
Output for equinox (D,M,Y) [20,5,1977]         ? 1,11,1977

Calculate position for date (D,M,Y) [20,5,1977] ? 1,11,1977
```

7900 PCOMET

```
Heliocentric coordinates: Comet Kohler (1977m)
------------ ------------

Ecliptic longitude (D,M,S) ...........    +341 24 52.90
Ecliptic latitude (D,M,S) ............    +  22  0 46.17
Radius vector (AU) ...................    1.004331

Geocentric coordinates: Comet Kohler (1977m)
---------- ------------

Ecliptic longitude (D,M,S) ...........    +276 34 27.22
Ecliptic latitude (D,M,S) ............    +  22 15 15.97
Distance from Earth (AU) .............    0.994116
Astrometric right ascension (H,M,S) ..       18 24 19.33
Astrometric declination (D,M,S) ......    -   1  2 50.28

Solar elongation (degrees) ...........    60.7

Again (Y or N) .......................    ? Y
New object (Y or N) ..................     ? N
New equinoxes (Y or N) ...............    ? Y

Equinox date of elements (D,M,Y) [0.9,1,1950]    ? ,,,
Output for equinox (D,M,Y) [1,11,1977]           ? 6,5,1978

Calculate position for date (D,M,Y) [1,11,1977]  ? 6,5,1978

Heliocentric coordinates: Comet Kohler (1977m)
------------ ------------

Ecliptic longitude (D,M,S) ...........    +  92 59 29.35
Ecliptic latitude (D,M,S) ............    -  48 42 47.91
Radius vector (AU) ...................    2.802193

Geocentric coordinates: Comet Kohler (1977m)
---------- ------------

Ecliptic longitude (D,M,S) ...........    +  76 29 12.68
Ecliptic latitude (D,M,S) ............    -  38 38  8.65
Distance from Earth (AU) .............    3.372313
Astrometric right ascension (H,M,S) ..        5 16 15.72
Astrometric declination (D,M,S) ......    -  15 42 26.35

Solar elongation (degrees) ...........    48.2

Again (Y or N) .......................    ? N
```

8100 PFIT
8800 EFIT

These routines (with their handling programs) calculate sets of orbital elements consistent with three or more observations of the position of a member of the Solar System made at different times. PFIT finds the parabolic elements; EFIT finds the elliptical elements.

The prediction of the position of a planet, comet, or any other member of the Solar System, depends upon knowing its orbital elements. The orbit is completely defined by these elements (in the absence of perturbations) and hence the position of the body can be calculated for any moment in the future or the past. Sometimes, however, you may not know the elements, particularly if you have been fortunate enough to have discovered a new comet. In that case, you can measure the position of the comet on at least three separate occasions as far apart as possible and then apply one of a number of methods to deduce the set of elements which are consistent with your observations. Unfortunately, the standard methods are neither particularly easy to apply, nor to understand, so I have adopted a more obvious, if perhaps rather flat-footed, approach to the problem which makes use of the mindless number-crunching ability of a computer.

The method is best understood by first considering a simpler problem in one dimension only. Suppose that we have a mathematical relationship between the variables VO and P such that VO is a function of P i.e. given any value of P we can immediately calculate the value of VO. Such a relationship could, for example, be Kepler's equation for an eccentricity of 1:

$$VO = P - SIN(P).$$

It is easy to calculate VO for any P with a calculator, but doing the problem the other way around is rather more difficult. If we know VO already but wish to find P, we discover that the equation cannot be solved directly and we have to adopt another approach such as that used in routine ANOMALY (3300).

The relationship between VO (*Visual Observations*) and P (*orbital Parameters*) in this hypothetical one-dimensional case might be as illustrated in Figure 8 by the solid curve. We are given a value of VO, marked VO_0, and the problem is to find the corresponding value of P_0. The method adopted by PFIT and EFIT works like this:

(i)   First, *guess* a value of P, say P_1.

(ii)   Calculate the corresponding VO_1.

217

(iii)   Find the slope (or derivative) of the curve at this point, DE_1.

(iv)   A better approximation to P is then

$$P_2 = P_1 + VC_1 \times (DE_1)^{-1},$$

where $VC_1 = (VO_0 - VO_1)$. Set P to this new value and go back to (i). Repeat the procedure as many times as necessary until the difference, VC, between the given value of VO and the latest approximation to it is sufficiently small.

The problem we actually face in finding the parabolic or elliptical orbital elements of a comet from N observations of its position is five-dimensional (parabolic) or six-dimensional (elliptical) so that the graph of Figure 8 must be replaced by an unimaginable one with six orthogonal axes. Mathematically, we deal with this by making P, VO, VC, and DE multi-dimensional *matrices*, rather than single-valued variables. We then have to solve the matrix equation

$$[\underline{P}(M)]_2 = [\underline{P}(M)]_1 + [\underline{VC}(2 \times N)]_1 \times [\underline{DE}(2 \times N, M)]_1^{-1}$$

using matrix algebra rather than common algebra, where M is 5 (parabolic) or 6 (elliptical), and $2 \times N$ is the number of independent observables (i.e. each observation provides two results, longitude and latitude). The problem is over-specified when there are more observables than orbital elements to be deduced from them. The reduction procedure used here is based on the method of conjugate gradients and it makes best use of all the observables to give the best-fitting solution. The measurements will usually not be exact because of experimental errors, but it is assumed that the intrinsic error in each is the same. I am indebted to Dr D. J. Thomas of Peterhouse, Cambridge for providing the matrix-manipulation algorithm.

One of the difficulties with this approach is that it tends to be unstable, the estimates of P leaping around the M-dimensional space almost at random unless the function is particularly well behaved. We can control it by limiting the maximum step allowed between one trial solution and the next so that P cannot change very fast. Also, we can make use of the fact that the parameters

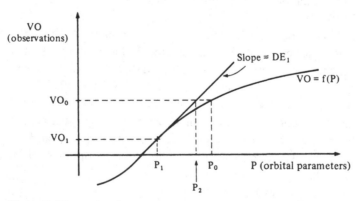

Figure 8. Illustrating the method of successive approximations used in PFIT and EFIT.

themselves are bounded. The argument of the perihelion, for example, must always be in the range 0–360 degrees. The perihelion distance and mean distance could, theoretically, take any values, but are limited here to the range 0.05–10 AU (perihelion distance), and greater than 0.1 AU (mean distance).

Routines PFIT and EFIT accept between three and six (inclusive) observations of the geocentric ecliptic coordinate pairs longitude and latitude in the array VO, made at the epochs given in array KP as the number of Julian days since 1900 January 0.5. They also require your initial guesses of the values of the elements. For PFIT, the elements are: KX, the epoch of the perihelion, QX, the perihelion distance, IX, the inclination, RX, the argument of the perihelion, and OX, the longitude of the ascending node. In the case of EFIT, the elements are: AX, the mean anomaly at the epoch, QX, the mean distance, IX, the inclination, RX, the argument of the perihelion, OX, the longitude of the ascending node, and EX, the eccentricity. You must also guess the epoch, KE, in Julian days since 1900 January 0.5.

The maximum step size is controlled by the step factor, SF. It takes a value in the range 0.1–1, depending on how close to the solution you are. As the absolute error between your latest estimates and the observations narrows, so the step factor increases, giving the routine more freedom to roam where it will.

Error conditions are reported by the error flags ER(10) for PFIT and ER(11) for EFIT. These normally have the value 0, but in certain circumstances the matrix manipulation may be impossible and then their values become 1. The message '** *intractable* **' is also printed. The handling program invites you to make another guess at the starting values of the orbital elements, although the problem may have arisen from an impossible set of observations through errors in the measurements or in entering the numbers into the program.

Each of the routines has two entry points, at lines 8100, 8145 for PFIT, and 8800, 8845 for EFIT. When the routines are called for the first time, it is necessary to calculate the position of the Earth at each of the observation times. These positions do not alter unless different observing times are used and hence subsequent calls can by-pass these calculations, saving a considerable amount of time. The first call, labelled pass 0, does not produce a new estimate of the orbital parameters. Having performed the Sun calculations, the routines return with the value of the absolute error, AF. You can then see quickly how good your first guess is, and amend it as you see fit (the handling programs give you this option).

Subsequent calls return with new estimates of the orbital parameters. The handling programs make repeated calls, dividing them up into groups called 'trials'. The first call of all is always to the first entry point where a parameter PA is set equal to −1. This parameter is incremented by the handling program after each return, and after the number of passes per trial (NT) the operator regains control. He or she then has the option of (i) continuing for another trial, (ii) entering new starting values, or (iii) stopping. A typical reduction requires a few trials of five passes each.

Progress towards the best solution is indicated by the variable AF, the absolute error. This is calculated just before control is returned to the calling program, and is a positive quantity which specifies the total error between each of the observed and calculated positions. It typically begins with a value of several hundred degrees but converges eventually to a minimum figure which depends upon the errors in the observations and the precision of the computer. In the examples shown later, exact values calculated by other programs in this book are used so that the ultimate errors are rather small. Note that the speed with which the iterations converge depends, amongst other things, upon the precision of your machine. You'll get much better results with greater precision; I have used eight-byte precision in the examples shown below.

When reducing a new set of observations, I suggest that you begin with the orbital elements in the middle of their respective ranges, and with the epoch set to your best guess. If a comet is approaching the Sun, for example, KX should be later than your last observation. The general trend of AF should be towards zero, albeit with many reverses. Be patient as it may take a long time to converge to the best solution. If you feel that progress is too slow, or if the program seems to get stuck, you can always try starting with a different set of values. I suggest that you practise by using PCOMET, for example, to produce a number of fictitious observations and then using PFIT to deduce the orbital parameters, comparing the best solution with the values you actually used in the first place. You'll find that the program works best with widely-separated observations, especially ones made either side of perihelion, becoming less accurate and requiring more passes as the observations move closer together.

The handling programs, HPFIT and HEFIT, make use of the input routine DEFAULT (880). See that section for details. Progress through the calculations, which may be quite slow, is indicated by the occasional display of an asterix, *, to reassure the operator that the program is still running!

Details of PFIT

Called by: GOSUB 8100 (first call for given observations), GOSUB 8145 (subsequent calls).

Converges on the best set of parabolic orbital elements consistent with a given set of NO% observations (3–6) of the position of a comet. The observations are input in the array VO, with alternate elements being set to geocentric ecliptic longitudes (*radians*; 1,3,5 etc.) and geocentric ecliptic latitudes (*radians*; 2,4,6 etc.). The epoch of the observation number J (1–NO%) is input via the array

KP(I,J), with I=1 for the day number (including the fraction), I=2 for the month number, I=3 for the year number, and I=4 for the corresponding number of Julian days since 1900 January 0.5.

The current estimates of the orbital parameters are input and output as KX, the epoch of the perihelion (Julian days since 1900 January 0.5), QX, the perihelion distance (*AU*), IX, the inclination (*radians*), RX, the argument of the perihelion (*radians*), and OX, the longitude of the ascending node (*radians*). The absolute error between the observed positions and those calculated with the current set of elements is output via AF (*radians*).

Arrays VO(12), DE(12,6), VC(12), LW(6), KP(4,6), EO(12), MW(6,6), DM(6), WS(6), TW(6), and WA(12) must be declared before calling the routine.

Error flag ER(10) is set to 1 if the matrix manipulation is impossible, and is set to 0 otherwise.

PA is set to -1 after a call to line 8100, but is unaffected by calls to line 8145. Other routine called: SUN (3400).

Details of EFIT

Called by: GOSUB 8800 (first call for given observations), GOSUB 8845 (subsequent calls).

Converges on the best set of elliptical orbital elements consistent with a given set of NO% observations (3–6) of the position of a celestial body. The observations are input in the array VO, with alternate elements being set to geocentric ecliptic longitudes (*radians*; 1,3,5 etc.) and geocentric ecliptic latitudes (*radians*; 2,4,6 etc.). The epoch of the observation number J (1–NO%) is input via the array KP(I,J), with I=1 for the day number (including the fraction), I=2 for the month number, I=3 for the year number, and I=4 for the corresponding number of Julian days since 1900 January 0.5.

The current estimates of the orbital parameters are input and output as AX, the mean anomaly of the epoch (*radians*), QX, the mean distance (*AU*), IX, the inclination (*radians*), RX, the argument of the perihelion (*radians*), OX, *the longitude of the ascending node (radians)*, and EX, the eccentricity. You must also specify KE, the epoch, in days since 1900 January 0.5. The absolute error between the observed positions and those calculated with the current set of elements is output via AF (*radians*).

Arrays VO(12), DE(12,6), VC(12), LW(6), KP(4,6), EO(12), MW(6,6), DM(6), WS(6), TW(6), and WA(12) must be declared before calling the routine.

Error flag ER(11) is set to 1 if the matrix manipulation is impossible, and is set to 0 otherwise.

8100 PFIT and 8800 EFIT

PA is set to -1 after a call to line 8800, but is unaffected by calls to line 8845. The maximum allowed values of QX and EX are limited by input parameters QM and EM respectively. Set these to, say QX=50, EX=0.95, before calling the routine.

Other routines called: ANOMALY (3300) and SUN (3400).

```
8097 REM
8098 REM       Subroutine PFIT
8099 REM
```

8100 UT=0, time carried as fraction of day; initialise PA

```
8100 PI=3.1415926535 : FL(1)=1 : UT=0 : PA=-1
```

8105 For each observation pair (NO% of them)...

```
8105 FOR I%=1 TO NO%
```

8110 ...get DJ from KP and call SUN...

8115 ...setting EH to Earth longitude and Sun–Earth distance

```
8110 DJ=KP(4,I%) : GOSUB 3400
8115 EH(1,I%)=SR+PI : EH(2,I%)=RR
8120 NEXT I%
```

8125 NN% observables; skip to absolute error calculation

```
8125 TP=2.0*PI : ER(10)=0 : NN%=2*NO% : GOTO 8455

8140 REM       Secondary entry point
```

8145 For each observable...

8150 ...call local routine to find the smallest...

8155 ...difference between observation and prediction

```
8145 FOR I%=1 TO NN%
8150 X=EO(I%) : Y=VO(I%) : GOSUB 8655 : VC(I%)=DX

8155 NEXT I%
```

8160 Find the gradients by calling local routines...

8165 ...in KX, epoch of perihelion...

```
8160 DV=2.0E-3 : DW=DV/2.0 : JL%=1 : KI=KX+DW

8165 GOSUB 8505 : KI=KX-DW : GOSUB 8520 : KI=KX
```

8170 ...in IX, inclination; use small angle approximations...

8175 ...for $\sin(x + dx)$ and $\cos(x + dx)$...

8180 ...to save execution time...

```
8170 W1=WI : W2=WJ : IN=IX+DW : WI=W1+DW*W2 : JL%=2

8175 WJ=W2-DW*W1 : GOSUB 8505 : IN=IX-DW : WI=W1-DW*W2
8180 WJ=W2+DW*W1 : GOSUB 8520 : IN=IX : WI=W1 : WJ=W2
```

8185 ...in OX, longitude of the ascending node...

```
8185 OM=OX+DW : GOSUB 8505 : OM=OX-DW
8190 JL%=3 : GOSUB 8520 : OM=OX
```

8195 ...in RX, argument of the perihelion...

```
8195 AR=RX+DW : GOSUB 8505 : AR=RX-DW
8200 JL%=4 : GOSUB 8520 : AR=RX
```

8205 ...and in QX, perihelion distance; use small increment...

8210 ...approximation for $(x + dx) \times$ $SQR(x + dx)$...

8215 ...to save execution time

```
8205 QP=QX+DW : QW=QQ : AA=1.5*DW/QX : QQ=QW*(1.0+AA)

8210 GOSUB 8505 : QP=QX-DW : QQ=QW*(1.0-AA)

8215 JL%=5 : GOSUB 8520 : QQ=QW : QP=QX
```

8220 Matrix-manipulation section; use integers...

8225 ...indicated by '%' wherever possible to make...

8230 ...the program quicker

```
8220 FOR JI%=1 TO 5 : LW(JI%)=0.0

8225 FOR JJ%=1 TO NN%

8230 LW(JI%)=LW(JI%)+DE(JJ%,JI%)*VC(JJ%)
8235 NEXT JJ%

8240 FOR JJ%=1 TO JI% : MW(JI%,JJ%)=0.0

8245 FOR JK%=1 TO NN%
8250 MW(JI%,JJ%)=MW(JI%,JJ%)+DE(JK%,JI%)*DE(JK%,JJ%)
8255 NEXT JK%

8260 MW(JJ%,JI%)=MW(JI%,JJ%)

8265 NEXT JJ%

8270 WS(JI%)=LW(JI%) : DM(JI%)=0

8275 NEXT JI%
8280 FOR JI%=1 TO 5                    ☞
```

```
8285 FOR JJ%=1 TO 5 : TW(JJ%)=0

8290 FOR JK%=1 TO 5
8295 TW(JJ%)=TW(JJ%)+MW(JJ%,JK%)*WS(JK%)
8300 NEXT JK%

8305 NEXT JJ%

8310 U=0.0 : W=0.0
8315 FOR JJ%=1 TO 5
8320 U=U+LW(JJ%)*WS(JJ%) : W=W+TW(JJ%)*WS(JJ%)
8325 NEXT JJ%
```

8330	. . . Matrix is singular if W=0	8330	`IF W<>0.0 THEN GOTO 8340`
8335	Error return from PFIT	8335	`ER(10)=1 : PRINT "** intractable **" : RETURN`

```
8340 U=U/W
8345 FOR JJ%=1 TO 5
8350 DM(JJ%)=DM(JJ%)+WS(JJ%)*U
8355 LW(JJ%)=LW(JJ%)-TW(JJ%)*U
8360 NEXT JJ%

8365 U=0.0 : W=0.0
8370 FOR JJ%=1 TO 5
8375 U=U+LW(JJ%)*TW(JJ%) : W=W+WS(JJ%)*TW(JJ%)
8380 NEXT JJ%
```

8385	Matrix is singular if W=0	8385	`IF W<>0.0 THEN GOTO 8395`
8390	Error return from PFIT	8390	`ER(10)=1 : PRINT "** intractable **" : RETURN`

```
8395 U=-U/W
8400 FOR JJ%=1 TO 5
8405 WS(JJ%)=LW(JJ%)+U*WS(JJ%)
8410 NEXT JJ%
```

8415	. . . end of matrix-manipulation section	`8415 NEXT JI%`
8420	Add increments to old estimates to obtain new ones . . .	`8420 KX=KX+SF*DM(1) : Y=TP`
8425	. . . reducing to primary ranges (e.g. $0-2\pi$) . . .	`8425 X=OX+SF*DM(3) : GOSUB 8640 : OX=X`
8430	. . . wherever possible; SF is current step factor . . .	`8430 X=RX+SF*DM(4) : GOSUB 8640 : RX=X`
8435	. . . in range 0.1–1.0	`8435 QX=QX+SF*DM(5)` `8440 X=IX+SF*DM(2) : Y=PI : GOSUB 8640 : IX=X`
8445	Limit range of QX to be greater than 0.05 . . .	`8445 IF QX<0.05 THEN QX=0.05`
8450	. . . but not greater than 10	`8450 IF QX>10.0 THEN QX=10.0`
8455	Set parameters to their current estimates . . .	`8455 KI=KX : IN=IX : OM=OX : AR=RX : QP=QX`
8460	. . . and calculate expected positions with them . . .	`8460 WI=SIN(IN) : WJ=COS(IN) : QQ=QP*SQR(QP)`
8465	. . . at each epoch of observation, by calling local routine	`8465 GOSUB 8535`
8470	Now find sum of the squares of the differences . . .	`8470 AF=0.0`
8475	. . . between expected and observed positions . . .	`8475 FOR I%=1 TO NN% STEP 2 : I1%=I1%+1`
8480	. . . at each of the epochs of observation . . .	`8480 X=VO(I%) : Y=EO(I%) : GOSUB 8655`

8485	allowing for geometrical factors . . .	`8485 A=DX*COS(EO(I1%)) : B=VO(I1%)-EO(I1%)`
		`8490 AF=AF+(A*A)+(B*B)`
		`8495 NEXT I%`

224

8500	. . . take the square root to obtain AF; normal return.	`8500 AF=SQR(AF) : PRINT : RETURN`
8505	First entry point to find gradient; find expected . . .	`8505 GOSUB 8535`
8510	. . . positions at each epoch of observation . . .	`8510 FOR I%=1 TO NN% : WA(I%)=EO(I%) : NEXT I%`
8515	. . . and save in WA	`8515 RETURN`
8520	Second entry point to find gradient; find expected . . .	`8520 GOSUB 8535 : FOR I%=1 TO NN%`
8525	. . . positions at each epoch of observation, get smallest . . .	`8525 X=EO(I%) : Y=WA(I%) : GOSUB 8655 : DE(I%,JL%)=DX/DV`
8530	. . . difference with previous values, and calculate gradient	`8530 NEXT I% : RETURN`
8535	Local routine to find expected position at each epoch . . .	`8535 FOR I%=1 TO NO% : IJ%=2*I% : II%=IJ%-1`
8540	. . . of observation. Call local routine for each epoch . . .	`8540 DJ=KP(4,I%) : LG=EH(1,I%) : RE=EH(2,I%) : GOSUB 8560`
8545	. . . and return position in EO	`8545 EO(II%)=EP : EO(IJ%)=BP` `8550 NEXT I%`
8555	Display a 'progress' asterisk	`8555 PRINT "*"; : RETURN`
8560	Local routine to find one position of comet . . .	`8560 W=((DJ-KI)*3.649116E-2)/QQ : S=W/3.0`
8565	. . . given the current estimates of the orbital parameters.	`8565 S2=S*S : D=(S2+3.0)*S-W`
8570	This routine is similar to PCOMET, but without . . .	`8570 IF ABS(D)<1E-6 THEN GOTO 8580`
8575	. . . allowance for light-travel time.	`8575 S=((2.0*S*S2)+W)/(3.0*(S2+1.0)) : GOTO 8565`
		`8580 NU=2.0*ATN(S) : R=QP*(1.0+S2) : L=NU+AR` `8585 S1=SIN(L) : C1=COS(L) : S2=S1*WI`
8590	FNS returns inverse sine	`8590 PS=FNS(S2) : LC=ATN(S1*WJ/C1)+OM : C2=COS(PS)`
8595	Resolve ambiguity of inverse tangent	`8595 IF C1<0 THEN LC=LC+PI` `8600 IF LC>TP THEN LC=LC-TP : GOTO 8600`
		`8605 RD=R*C2 : LL=LC-LG : C3=COS(LL) : S3=SIN(LL)`
8610	Inner comet or outer comet?	`8610 IF RD<RE THEN GOTO 8620`
8615	Outer . . .	`8615 EP=ATN((RE*S3)/(RD-(RE*C3)))+LC : GOTO 8625`
8620	. . . inner	`8620 EP=ATN((-RD*S3)/(RE-(RD*C3)))+LG+PI`
8625	Restore to interval 0–2π	`8625 Y=TP : X=EP : GOSUB 8640 : EP=X` `8630 TB=(RD*S2*SIN(EP-LC))/(C2*RE*S3)`
8635	Return with position in EP, BP	`8635 BP=ATN(TB) : RETURN`

8640	Restore X to the range 0–Y	`8640 IF X<0 THEN X=X+Y : GOTO 8640`
		`8645 IF X>Y THEN X=X-Y : GOTO 8645`
		`8650 RETURN`

8655	Routine to return the smallest difference, DX, . . .	`8655 DX=X-Y : A=ABS(DX)`
8600	. . . between X and Y, both being assumed to be . . .	`8660 IF A<PI THEN RETURN`

8665	. . . in the range 0–2π	`8665 B=DX+TP : C=DX-TP`
		`8670 IF ABS(B)<A THEN DX=B : RETURN`
		`8675 DX=C : RETURN`

Notes		Code
	1	REM
	2	REM Handling program HPFIT
	3	REM
5 Declare the arrays	5	DIM FL(20),ER(20),SW(20)
	10	DIM VO(12),DE(12,6),VC(12),LW(6),KP(4,6),EO(12)
	15	DIM MW(6,6),DM(6),WS(6),TW(6),WA(12),EH(2,6)
20 Truncated-integer function	20	DEF FNI(W)=SGN(W)*INT(ABS(W))
25 Least-integer function	25	DEF FNL(W)=FNI(W)+FNI((SGN(W)-1.0)/2.0)
30 Returns value of W in string format, to six decimal places	30	DEF FNQ$(W)=STR$(INT(W*1E6+0.5)/1E6)
35 Converts from radians to degrees . . .	35	DEF FNM(W)=1.745329252E-2*W
40 . . . and vice-versa	40	DEF FND(W)=5.729577951E1*W
45 Returns inverse sine	45	DEF FNS(W)=ATN(W/(SQR(1-W*W)+1E-20))
	50	PRINT : PRINT "Fitting parabolic elements"
	55	PRINT "----------------------------" : PRINT : PRINT
60 Initial default values	60	QX=2.0 : IX=45 : RX=180 : OX=180
	65	NT=5
70 Use DEFAULT for all inputs	70	PRINT
75 Get number of observations NO% . . .	75	QT$="How many observation to fit (3-6)" : Q$=QT$
	80	X=NO% : N=1 : GOSUB 880 : NO%=X
85 . . . minimum 3, maximum 6	85	IF X<3.0 OR X>6.0 THEN PRINT "What ? " : GOTO 75
90 Now get the observations of the comet . . .	90	Q$="Please input the observations:"
	95	PRINT : PRINT Q$
	100	QA$="Observation number"
	105	QB$="Epoch of observation (D,M,Y)"
	110	QC$="Ecliptic longitude (degrees)"
	115	QD$="Ecliptic latitude (degrees)"
120 . . . loop round for NO% times . . .	120	FOR I%=1 TO NO% : IJ%=2*I% : II%=IJ%-1
	125	PRINT : PRINT QA$; I% : Q$=QB$: X=KP(1,I%)
130 . . . calendar date of epoch . . .	130	Y=KP(2,I%) : Z=KP(3,I%) : N=3 : GOSUB 880
	135	KP(1,I%)=X : KP(2,I%)=Y : KP(3,I%)=Z
140 . . . ecliptic longitude . . .	140	Q$=QC$: X=VO(II%) : N=1 : GOSUB 880 : VO(II%)=X
145 . . . and ecliptic latitude	145	Q$=QD$: X=VO(IJ%) : N=1 : GOSUB 880 : VO(IJ%)=X
	150	NEXT I%
155 Better check that we've got them right . . .	155	PRINT : PRINT "Your values are"
160 . . . display the inputs . . .	160	PRINT "---------------" : PRINT
	165	FOR I%=1 TO NO% : IJ%=2*I% : II%=IJ%-1
	170	PRINT QA$; I%
	175	PRINT QB$; TAB(52); FNQ$(KP(1,I%));
	180	PRINT INT(KP(2,I%)); INT(KP(3,I%))
	185	PRINT QC$; TAB(52); FNQ$(VO(II%))
	190	PRINT QD$; TAB(52); FNQ$(VO(IJ%))
	195	NEXT I%
200 . . . and find out if OK . . .	200	Q$="Is this correct (Y or N) "
205 . . . with YESNO	205	PRINT : GOSUB 960 : PRINT
210 Return to base if any errors	210	IF E=0 THEN GOTO 70
215 EE is local error flag	215	EE=0
220 For each observation . . .	220	FOR I%=1 TO NO%
	225	DY=KP(1,I%) : MN=KP(2,I%) : YR=KP(3,I%)
230 . . . convert calendar date to Julian form with JULDAY . . .	230	FL(1)=0 : GOSUB 1100 : EE=EE+ER(1) : KP(4,I%)=DJ
	235	NEXT I%
240 . . . and return to base if any date is impossible	240	IF EE<>0 THEN GOTO 70

☛

245	NN% is number of observables (two per observation)...	245	`NN%=2*NO%`
250	... i.e. longitude and latitude; convert to radians	250	`FOR I%=1 TO NN% : VO(I%)=FNM(VO(I%)) : NEXT I%`
255	Now for your guesses...	255	`Q$="Please input starting values:"`
		260	`PRINT : PRINT Q$ : PRINT`
265	... get the values...	265	`QF$="Epoch of perihelion (D,M,Y)" : Q$=QF$`
		270	`X=DI : Y=MI : Z=YI : N=3 : GOSUB 880`
		275	`DI=X : MI=Y : YI=Z`
		280	`QG$="Perihelion distance (AU)" : Q$=QG$`
		285	`X=QX : N=1 : GOSUB 880 : QX=X`
		290	`QH$="Inclination of orbit (degrees)" : Q$=QH$`
		295	`X=IX : N=1 : GOSUB 880 : IX=X`
		300	`QI$="Argument of perihelion (degrees)" : Q$=QI$`
		305	`X=RX : N=1 : GOSUB 880 : RX=X`
		310	`QJ$="Longitude of ascending node (deg.)" : Q$=QJ$`
		315	`X=OX : N=1 : GOSUB 880 : OX=X`
		320	`QK$="How many passes per trial" : Q$=QK$`
		325	`X=NT : N=1 : GOSUB 880 : NT=X : PRINT`
330	... and display them to check ok.	330	`PRINT "Your values are"`
		335	`PRINT "--------------" : PRINT`
		340	`PRINT QF$; TAB(52); FNQ$(DI); INT(MI); INT(YI)`
		345	`PRINT QG$; TAB(52); FNQ$(QX)`
		350	`PRINT QH$; TAB(52); FNQ$(IX)`
		355	`PRINT QI$; TAB(52); FNQ$(RX)`
		360	`PRINT QJ$; TAB(52); FNQ$(OX)`
		365	`PRINT QK$; TAB(52); FNQ$(NT)`
370	Find out if correct...	370	`Q$="Is this correct (Y or N) "`
375	... by calling YESNO	375	`PRINT : GOSUB 960 : PRINT`
380	Get some more starting values if not ok	380	`IF E=0 THEN GOTO 255`
385	Convert to radians, and to Julian date form...	385	`IX=FNM(IX) : RX=FNM(RX) : OX=FNM(OX) : DY=DI`
390	... by calling JULDAY	390	`MN=MI : YR=YI : FL(1)=0 : GOSUB 1100 : KX=DJ`
395	Impossible date?	395	`IF ER(1)<>0 THEN GOTO 255`
400	Initialise these strings	400	`QN$="Absolute error (degrees)" : QM$="Step factor"`
405	Set step factor to 0 (irrelevant) and call PFIT...	405	`SF=0.0 : GOSUB 8100`
410	... at initial entry point; ER(10) should be 0	410	`IF ER(10)=1 THEN GOTO 60`
415	Increment pass counter...	415	`PA=PA+1 : PRINT`
420	... and display the results...	420	`PRINT "Results after pass "; PA : PRINT`
425	... epoch of peri. in calendar form with CALDAY	425	`DJ=KX : FL(2)=0 : GOSUB 1200`
		430	`PRINT QF$; TAB(52); FNQ$(DY); INT(MN); INT(YR)`
		435	`PRINT QG$; TAB(52); FNQ$(QX)`
		440	`PRINT QH$; TAB(52); FNQ$(FND(IX))`
		445	`PRINT QI$; TAB(52); FNQ$(FND(RX))`
		450	`PRINT QJ$; TAB(52); FNQ$(FND(OX))`
		455	`PRINT QM$; TAB(52); FNQ$(SF)`
		460	`PRINT QN$; TAB(52); FNQ$(FND(AF))`
465	Set step factor according to value of absolute error	465	`IF FND(AF)>180.0 THEN SF=0.1 : GOTO 490`
		470	`IF FND(AF)>135.0 THEN SF=0.2 : GOTO 490`
		475	`IF FND(AF)>90.0 THEN SF=0.3 : GOTO 490`
		480	`IF FND(AF)>45.0 THEN SF=0.5 : GOTO 490`
		485	`SF=1.0`
490	Have we completed a trial?	490	`P1=PA-INT(PA/NT)*NT`
495	If not, call PFIT at secondary point, and loop	495	`IF P1<>0 THEN PRINT : GOSUB 8145 : GOTO 415`
500	Trial completed; what next?	500	`Q$="Continue, new values, stop (C/N/S)"`
		505	`PRINT : PRINT Q$; : INPUT A$`

```
510   IF A$="S" OR A$="s" THEN STOP
515   IF A$="C" OR A$="c" THEN GOTO 530
520   IF A$="N" OR A$="n" THEN GOTO 535
```

525 Answer not recognised; get another one

```
525   PRINT "What ? " : GOTO 500
```

530 Carry on with the next trial

```
530   PRINT : GOSUB 8145 : GOTO 415
```

535 New values requires; use old values as defaults . . .

```
535   NN%=2*NO%
```

540 . . . converting first to degrees

```
540   FOR I%=1 TO NN% : VO(I%)=FND(VO(I%)) : NEXT I%
545   DI=DY : MI=MN : YI=YR : IX=FND(IX)
550   RX=FND(RX) : OX=FND(OX) : PRINT : GOTO 70
```

```
INCLUDE DEFAULT, YESNO, JULDAY, CALDAY, ANOMALY, SUN, PFIT
```

Example

```
Fitting parabolic elements
--------------------------


How many observation to fit (3-6) [0]          ? 3

Please input the observations:

Observation number 1
Epoch of observation (D,M,Y) [0,0,0]           ? 20,5,1977
Ecliptic longitude (degrees) [0]               ? 217.661106
Ecliptic latitude (degrees) [0]                ? 55.800497

Observation number 2
Epoch of observation (D,M,Y) [0,0,0]           ? 1,11,1977
Ecliptic longitude (degrees) [0]               ? 276.574228
Ecliptic latitude (degrees) [0]                ? 22.254436

Observation number 3
Epoch of observation (D,M,Y) [0,0,0]           ? 6,5,1978
Ecliptic longitude (degrees) [0]               ? 76.486856
Ecliptic latitude (degrees) [0]                ? -38.635736

Your values are
---------------

Observation number 1
Epoch of observation (D,M,Y)                   20  5   1977
Ecliptic longitude (degrees)                   217.661106
Ecliptic latitude (degrees)                    55.800497
Observation number 2
Epoch of observation (D,M,Y)                   1  11   1977
Ecliptic longitude (degrees)                   276.574228
Ecliptic latitude (degrees)                    22.254436
Observation number 3
Epoch of observation (D,M,Y)                   6  5   1978
Ecliptic longitude (degrees)                   76.486856
Ecliptic latitude (degrees)                    -38.635735

Is this correct (Y or N) ? Y
```

228

```
Please input starting values:

Epoch of perihelion (D,M,Y) [0,0,0]        ? 1,11,1977
Perihelion distance (AU) [2]               ? 1.5
Inclination of orbit (degrees) [45]        ? ,
Argument of perihelion (degrees) [180]     ? ,
Longitude of ascending node (deg.) [180]   ? ,
How many passes per trial [5]              ? ,

Your values are
----------------

Epoch of perihelion (D,M,Y)                1 11  1977
Perihelion distance (AU)                   1.5
Inclination of orbit (degrees)             45
Argument of perihelion (degrees)           180
Longitude of ascending node (deg.)         180
How many passes per trial                  5

Is this correct (Y or N) ? Y

*


Results after pass  0

Epoch of perihelion (D,M,Y)                1 11  1977
Perihelion distance (AU)                   1.5
Inclination of orbit (degrees)             45
Argument of perihelion (degrees)           180
Longitude of ascending node (deg.)         180
Step factor                                0
Absolute error (degrees)                   73.273461

Continue, new values, stop (C/N/S)? C

***********

Results after pass  1

Epoch of perihelion (D,M,Y)                1.971898 11   1977
Perihelion distance (AU)                   1.19699
Inclination of orbit (degrees)             36.065819
Argument of perihelion (degrees)           171.079493
Longitude of ascending node (deg.)         177.89961
Step factor                                0.5
Absolute error (degrees)                   37.263933

***********

Results after pass  2

Epoch of perihelion (D,M,Y)                7.6628 11   1977
Perihelion distance (AU)                   0.954606
Inclination of orbit (degrees)             43.928317
Argument of perihelion (degrees)           157.248253
Longitude of ascending node (deg.)         186.982385
Step factor                                1
Absolute error (degrees)                   10.124592

***********
```

☞

8100 PFIT and 8800 EFIT

```
Results after pass   3

Epoch of perihelion (D,M,Y)                    10.054192 11   1977
Perihelion distance (AU)                       0.987838
Inclination of orbit (degrees)                 48.611255
Argument of perihelion (degrees)               163.490038
Longitude of ascending node (deg.)             182.108532
Step factor                                    1
Absolute error (degrees)                       0.466675

***********

Results after pass   4

Epoch of perihelion (D,M,Y)                    10.562671 11   1977
Perihelion distance (AU)                       0.990599
Inclination of orbit (degrees)                 48.716622
Argument of perihelion (degrees)               163.475816
Longitude of ascending node (deg.)             182.205059
Step factor                                    1
Absolute error (degrees)                       4.61E-03

***********

Results after pass   5

Epoch of perihelion (D,M,Y)                    10.563567 11   1977
Perihelion distance (AU)                       0.990595
Inclination of orbit (degrees)                 48.718114
Argument of perihelion (degrees)               163.474024
Longitude of ascending node (deg.)             182.208533
Step factor                                    1
Absolute error (degrees)                       3.101E-03

Continue, new values, stop (C/N/S)? S
```

The elements converged upon represent best values for the equinox of (average) date. Use RELEM to convert to another equinox if necessary before comparing with the original values.

```
                              8797 REM
                              8798 REM        Subroutine EFIT
                              8799 REM
```

8800	UT=0, time carried as fraction of day; initialise PA
```
8800 PI=3.1415926535 : FL(1)=1 : UT=0 : PA=-1
```

8805	For each observation pair (NO% of them) . . .
```
8805 FOR I%=1 TO NO%
```

8810	. . . get DJ from KP and call SUN . . .
8815	. . . setting EH to Earth longitude and Sun–Earth distance
```
8810 DJ=KP(4,I%) : GOSUB 3400
8815 EH(1,I%)=SR+PI : EH(2,I%)=RR
8820 NEXT I%
```

8825	NN% observables; skip to absolute error calculation
```
8825 TP=2.0*PI : ER(11)=0 : NN%=2*NO% : GOTO 9170

8840 REM        Secondary entry point
```

8845	For each observable . . .
8850	. . . call local routine to find the smallest . . .
8855	. . . difference between observation and prediction
```
8845 FOR I%=1 TO NN%
8850 X=EO(I%) : Y=VO(I%) : GOSUB 9370 : VC(I%)=DX

8855 NEXT I%
```

8860	Find the gradients by calling local routines . . .
8865	. . . in AX, mean anomaly at the epoch. . .
```
8860 DV=2.0E-3 : DW=DV/2.0 : JL%=1 : AI=AX+DW

8865 GOSUB 9215 : AI=AX-DW : GOSUB 9230 : AI=AX
```

8870	. . . in IX, inclination; use small angle approximations . . .
8875	. . . for $\sin(x + dx)$ and $\cos(x + dx)$ . . .
8880	. . . to save execution time . . .
```
8870 W1=WI : W2=WJ : IN=IX+DW : WI=W1+DW*W2 : JL%=2
8875 WJ=W2-DW*W1 : GOSUB 9215 : IN=IX-DW : WI=W1-DW*W2
8880 WJ=W2+DW*W1 : GOSUB 9230 : IN=IX : WI=W1 : WJ=W2
```

8885	. . . in OX, longitude of the ascending node . . .
```
8885 OM=OX+DW : GOSUB 9215 : OM=OX-DW
8890 JL%=3 : GOSUB 9230 : OM=OX
```

8895	. . . in RX, argument of the perihelion . . .
```
8895 AR=RX+DW : GOSUB 9215 : AR=RX-DW
8900 JL%=4 : GOSUB 9230 : AR=RX
```

8905	. . . in QX, mean distance; use small increment . . .
```
8905 QP=QX+DW : GOSUB 9215 : QP=QX-DW
8910 JL%=5 : GOSUB 9230 : QP=QX
```

8915	. . . and in EC, the eccentricity
```
8915 EC=EX+DW : GOSUB 9215 : EC=EX-DW
8920 JL%=6 : GOSUB 9230 : EC=EX
```

8925	Matrix-manipulation section; use integers . . .
8930	. . . indicated by '%' wherever possible to make . . .
8935	. . . the program quicker
```
8925 FOR JI%=1 TO 6 : LW(JI%)=0.0

8930 FOR JJ%=1 TO NN%

8935 LW(JI%)=LW(JI%)+DE(JJ%,JI%)*VC(JJ%)
8940 NEXT JJ%

8945 FOR JJ%=1 TO JI% : MW(JI%,JJ%)=0.0

8950 FOR JK%=1 TO NN%
8955 MW(JI%,JJ%)=MW(JI%,JJ%)+DE(JK%,JI%)*DE(JK%,JJ%)
8960 NEXT JK%

8965 MW(JJ%,JI%)=MW(JI%,JJ%)

8970 NEXT JJ%

8975 WS(JI%)=LW(JI%) : DM(JI%)=0
8980 NEXT JI%

8985 FOR JI%=1 TO 6
```

8100 PFIT and 8800 EFIT

```
8990 FOR JJ%=1 TO 6 : TW(JJ%)=0

8995 FOR JK%=1 TO 6
9000 TW(JJ%)=TW(JJ%)+MW(JJ%,JK%)*WS(JK%)
9005 NEXT JK%

9010 NEXT JJ%

9015 U=0.0 : W=0.0
9020 FOR JJ%=1 TO 6
9025 U=U+LW(JJ%)*WS(JJ%) : W=W+TW(JJ%)*WS(JJ%)
9030 NEXT JJ%
```

9035 Matrix is singular if W=0
9040 Error return from EFIT

```
9035 IF W<>0.0 THEN GOTO 9045
9040 ER(11)=1 : PRINT "** intractable **" : RETURN

9045 U=U/W
9050 FOR JJ%=1 TO 6
9055 DM(JJ%)=DM(JJ%)+WS(JJ%)*U
9060 LW(JJ%)=LW(JJ%)-TW(JJ%)*U
9065 NEXT JJ%

9070 U=0.0 : W=0.0
9075 FOR JJ%=1 TO 6
9080 U=U+LW(JJ%)*TW(JJ%) : W=W+WS(JJ%)*TW(JJ%)
9085 NEXT JJ%
```

9090 Matrix is singular if W=0
9095 Error return from EFIT

```
9090 IF W<>0.0 THEN GOTO 9100
9095 ER(11)=1 : PRINT "** intractable **" : RETURN

9100 U=-U/W
9105 FOR JJ%=1 TO 6
9110 WS(JJ%)=LW(JJ%)+U*WS(JJ%)
9115 NEXT JJ%
```

9120 ...end of matrix-manipulation section

```
9120 NEXT JI%
```

9125 Add increments to old estimates to obtain new ones...

```
9125 AX=AX+SF*DM(1) : QX=QX+SF*DM(5)
```

9130 ...reducing to primary ranges (e.g. 0–2π)...

```
9130 EX=EX+SF*DM(6) : Y=PI
```

9135 ...wherever possible; SF is current step factor...

```
9135 X=IX+SF*DM(2) : GOSUB 9355 : IX=X : Y=TP
```

9140 ...in range 0.1–1.0

```
9140 X=OX+SF*DM(3) : GOSUB 9355 : OX=X
9145 X=RX+SF*DM(4) : GOSUB 9355 : RX=X
```

9150 Limit range of QX to be greater than 0.1...

```
9150 IF QX<0.1 THEN QX=0.1
```

9155 ...but not greater than QM (=50; set by handling program)

```
9155 IF QX>QM THEN QX=QM
```

9160 Limit EC to be in range...
9165 ...IE–3 (DW) to EM (=0.95; set by handling program)

```
9160 IF EX<DW THEN EX=DW
9165 IF EX>EM THEN EX=EM
```

9170 Set parameters to their current estimates...

```
9170 AI=AX : IN=IX : OM=OX : AR=RX : QP=QX : EC=EX
```

9175 ...and calculate expected positions with them

```
9175 WI=SIN(IN) : WJ=COS(IN) : GOSUB 9260
```

9180 Now find sum of the squares of the differences...

```
9180 AF=0.0
```

9185 ...between expected and observed positions...

```
9185 FOR I%=1 TO NN% STEP 2 : I1%=I1%+1
```

9190 ...at each of the epochs of observation...

```
9190 X=VO(I%) : Y=EO(I%) : GOSUB 9370
```

9195 ...allowing for geometrical factors...

```
9195 A=DX*COS(EO(I1%)) : B=VO(I1%)-EO(I1%)
9200 AF=AF+(A*A)+(B*B)
9205 NEXT I%
```

9210	...take the square root to obtain AF; normal return.	`9210 AF=SQR(AF) : PRINT : RETURN`
9215	...positions at each epoch of observation...	`9215 GOSUB 9260` `9220 FOR I%=1 TO NN% : WA(I%)=EO(I%) : NEXT I%`
9225	...and save in WA	`9225 RETURN`
9230	Second entry point to find gradient; find expected...	`9230 GOSUB 9260`
9235	...positions at each epoch of observation, get smallest...	`9235 FOR I%=1 TO NN%`

9240	...difference with previous values, and calculate gradient	`9240 X=EO(I%) : Y=WA(I%)`
		`9245 GOSUB 9370 : DE(I%,JL%)=DX/DV`
		`9250 NEXT I%`
		`9255 RETURN`

9260	Local routine to find expected position at each epoch...	`9260 ND=1.72027912E-2/SQR(QP*QP*QP) : AA=QP*(1.0-EC*EC)`
9265	...of observation. Call local routine for each epoch...	`9265 FOR I%=1 TO NO% : IJ%=2*I% : II%=IJ%-1`

9270	...and return position in EO	`9270 DJ=KP(4,I%) : LG=EH(1,I%) : RE=EH(2,I%) : GOSUB 9290`
		`9275 EO(II%)=EP : EO(IJ%)=BP`
		`9280 NEXT I%`

9285	Display a 'progress' asterisk	`9285 PRINT "*"; : RETURN`
9290	Local routine to find one position of object...	`9290 AM=AI+ND*(DJ-KE) : GOSUB 3300 : PV=AA/(1.0+EC*COS(AT))`
9295	...given the current estimates of the orbital parameters	`9295 LO=AT+AR : SO=SIN(LO) : CO=COS(LO) : SP=SO*WI` `9300 Y=SO*WJ : PD=ATN(Y/CO)+OM`
9305	This routine is similar to ELOSC, but without...	`9305 IF CO<0.0 THEN PD=PD+PI`
9310	...allowance for light travel time.	`9310 IF PD>TP THEN PD=PD-TP`
		`9315 CI=SQR(1.0-SP*SP) : RD=PV*CI : LL=PD-LG` `9320 L1=SIN(LL) : L2=COS(LL)`
9325	Inner object or outer object?	`9325 IF PV<RE THEN GOTO 9335`
9330	Outer...	`9330 EP=ATN(RE*L1/(RD-RE*L2))+PD : GOTO 9340`
9335	...inner	`9335 EP=ATN(-RD*L1/(RE-RD*L2))+LG+PI`
9340	Restore to interval 0–2π	`9340 Y=TP : X=EP : GOSUB 9355 : EP=X` `9345 TB=RD*SP*SIN(EP-PD)/(CI*RE*L1)`
9350	Return with position in EP, BP	`9350 BP=ATN(TB) : RETURN`

9355	Restore X to the range 0–Y	`9355 IF X<0 THEN X=X+Y : GOTO 9355`
		`9360 IF X>Y THEN X=X-Y : GOTO 9360`
		`9365 RETURN`

9370	Routine to return the smallest difference, DX,...	`9370 DX=X-Y : A=ABS(DX)`
9375	...between X and Y, both being assumed to be...	`9375 IF A<PI THEN RETURN`

9380	...in the range 0–2π	`9380 B=DX+TP : C=DX-TP`
		`9385 IF ABS(B)<A THEN DX=B : RETURN`
		`9390 DX=C : RETURN`

Note	Description	Line	Code
		1	REM
		2	REM Handling program HEFIT
		3	REM
5	Declare the arrays	5	DIM FL(20),ER(20),SW(20)
		10	DIM VO(12),DE(12,6),VC(12),LW(6),KP(4,6),EO(12)
		15	DIM MW(6,6),DM(6),WS(6),TW(6),WA(12),EH(2,6)
20	Truncated-integer function	20	DEF FNI(W)=SGN(W)*INT(ABS(W))
25	Least-integer function	25	DEF FNL(W)=FNI(W)+FNI((SGN(W)-1.0)/2.0)
30	Returns value of W in string format, to six decimal places	30	DEF FNQ$(W)=STR$(INT(W*1E6+0.5)/1E6)
35	Converts from radians to degrees ...	35	DEF FNM(W)=1.745329252E-2*W
40	... and vice-versa	40	DEF FND(W)=5.729577951E1*W
45	Returns inverse sine	45	DEF FNS(W)=ATN(W/(SQR(1-W*W)+1E-20))
		50	PRINT : PRINT "Fitting elliptical elements"
		55	PRINT "-------------------------" : PRINT : PRINT
60	Initial default values	60	QX=3.0 : IX=30 : RX=0 : OX=180 : EX=0.2
		65	NT=5
70	Use DEFAULT for all inputs	70	PRINT
75	Get number of observations NO% ...	75	QT$="How many observation to fit (3-6)" : Q$=QT$
		80	X=NO% : N=1 : GOSUB 880 : NO%=X
85	... minimum 3, maximum 6	85	IF X<3.0 OR X>6.0 THEN PRINT "What ? " : GOTO 75
90	Now get the observations of the object ...	90	Q$="Please input the observations:"
		95	PRINT : PRINT Q$
		100	QA$="Observation number"
		105	QB$="Epoch of observation (D,M,Y)"
		110	QC$="Ecliptic longitude (degrees)"
		115	QD$="Ecliptic latitude (degrees)"
120	... loop round for NO% times ...	120	FOR I%=1 TO NO% : IJ%=2*I% : II%=IJ%-1
		125	PRINT : PRINT QA$; I% : Q$=QB$: X=KP(1,I%)
130	... calendar date of epoch ...	130	Y=KP(2,I%) : Z=KP(3,I%) : N=3 : GOSUB 880
		135	KP(1,I%)=X : KP(2,I%)=Y : KP(3,I%)=Z
140	... ecliptic longitude ...	140	Q$=QC$: X=VO(II%) : N=1 : GOSUB 880 : VO(II%)=X
145	... and ecliptic latitude	145	Q$=QD$: X=VO(IJ%) : N=1 : GOSUB 880 : VO(IJ%)=X
		150	NEXT I%
155	Better check that we've got them right ...	155	PRINT : PRINT "Your values are"
160	... display the inputs ...	160	PRINT "---------------" : PRINT
		165	FOR I%=1 TO NO% : IJ%=2*I% : II%=IJ%-1
		170	PRINT QA$; I%
		175	PRINT QB$; TAB(52); FNQ$(KP(1,I%));
		180	PRINT INT(KP(2,I%)); INT(KP(3,I%))
		185	PRINT QC$; TAB(52); FNQ$(VO(II%))
		190	PRINT QD$; TAB(52); FNQ$(VO(IJ%))
		195	NEXT I%
200	... and find out if OK ...	200	Q$="Is this correct (Y or N) "
205	... with YESNO	205	PRINT : GOSUB 960 : PRINT
210	Return to base if any errors	210	IF E=0 THEN GOTO 70
215	EE is local error flag	215	EE=0
220	For each observation ...	220	FOR I%=1 TO NO%
		225	DY=KP(1,I%) : MN=KP(2,I%) : YR=KP(3,I%)
230	... convert calendar date to Julian form with JULDAY ...	230	FL(1)=0 : GOSUB 1100 : EE=EE+ER(1) : KP(4,I%)=DJ
		235	NEXT I%
240	... and return to base if any date is impossible	240	IF EE<>0 THEN GOTO 70

245	NN% is number of observables (two per observation) . . .	245	`NN%=2*NO%`
250	. . . i.e. longitude and latitude; convert to radians	250	`FOR I%=1 TO NN% : VO(I%)=FNM(VO(I%)) : NEXT I%`
255	Get the epoch for which the elements apply . . .	255	`QQ$="Epoch for elements (D,M,Y)" : Q$=QQ$`
260	. . . e.g. argument of perihelion at this epoch	260	`X=DE : Y=ME : Z=YE : N=3 : GOSUB 880`
		265	`DE=X : ME=Y : YE=Z`
270	Now for your guesses . . .	270	`Q$="Please input starting values:"`
		275	`PRINT : PRINT Q$ : PRINT`
280	. . . get the values . . .	280	`QF$="Mean anomaly at the epoch (deg.)" : Q$=QF$`
		285	`X=AX : N=1 : GOSUB 880 : AX=X`
		290	`QG$="Mean distance (AU)" : Q$=QG$`
		295	`X=QX : N=1 : GOSUB 880 : QX=X`
		300	`QH$="Inclination of orbit (degrees)" : Q$=QH$`
		305	`X=IX : N=1 : GOSUB 880 : IX=X`
		310	`QI$="Argument of perihelion (degrees)" : Q$=QI$`
		315	`X=RX : N=1 : GOSUB 880 : RX=X`
		320	`QJ$="Longitude of ascending node (deg.)" : Q$=QJ$`
		325	`X=OX : N=1 : GOSUB 880 : OX=X`
		330	`QO$="Eccentricity (0 to 1)" : Q$=QO$`
		335	`X=EX : N=1 : GOSUB 880 : EX=X`
		340	`IF X<0 OR X>1.0 THEN PRINT "What ?" : GOTO 335`
		345	`QK$="How many passes per trial" : Q$=QK$`
		350	`X=NT : N=1 : GOSUB 880 : NT=X : PRINT`
360	. . . and display them to check ok.	360	`PRINT "Your values are"`
		365	`PRINT "---------------" : PRINT`
		370	`PRINT QQ$; TAB(52); FNQ$(DE); INT(ME); INT(YE)`
		375	`PRINT QF$; TAB(52); FNQ$(AX)`
		380	`PRINT QG$; TAB(52); FNQ$(QX)`
		385	`PRINT QH$; TAB(52); FNQ$(IX)`
		390	`PRINT QI$; TAB(52); FNQ$(RX)`
		395	`PRINT QJ$; TAB(52); FNQ$(OX)`
		400	`PRINT QO$; TAB(52); FNQ$(EX)`
		405	`PRINT QK$; TAB(51); INT(NT)`
410	Find out if correct . . .	410	`Q$="Is this correct (Y or N) "`
415	. . . by calling YESNO	415	`PRINT : GOSUB 960 : PRINT`
420	Get some more starting values if not ok	420	`IF E=0 THEN GOTO 255`
425	Convert to radians, and to Julian date form . . .	425	`IX=FNM(IX) : RX=FNM(RX)`
430	. . . by calling JULDAY	430	`OX=FNM(OX) : AX=FNM(AX)`
		435	`DY=DE : MN=ME : YR=YE : FL(1)=0`
		440	`GOSUB 1100 : KE=DJ`
445	Impossible date?	445	`IF ER(1)<>0 THEN GOTO 255`
450	Initialise these strings	450	`QN$="Absolute error (degrees)"`
		455	`QM$="Step factor"`
		460	`QS$="Mean daily motion (degrees)"`
465	Set limits on QX and EX and call EFIT . . .	465	`QM=50.0 : EM=0.95 : SF=0.0 : GOSUB 8800`
470	. . . at initial entry point; ER(11) should be 0	470	`IF ER(11)=1 THEN GOTO 60`
475	Increment pass counter . . .	475	`PA=PA+1 : PRINT`
480	. . . and display the results . . .	480	`PRINT "Results after pass "; PA : PRINT`
		485	`ND=0.985647332/SQR(QX*QX*QX)`
		490	`PRINT QS$; TAB(52); FNQ$(ND)`
		495	`PRINT QF$; TAB(52); FNQ$(FND(AX))`
		500	`PRINT QG$; TAB(52); FNQ$(QX)`
		505	`PRINT QH$; TAB(52); FNQ$(FND(IX))`
		510	`PRINT QI$; TAB(52); FNQ$(FND(RX))`
		515	`PRINT QJ$; TAB(52); FNQ$(FND(OX))`

```
                        520   PRINT QO$; TAB(52); FNQ$(EX)
                        525   PRINT QM$; TAB(52); FNQ$(SF)
                        530   PRINT QN$; TAB(52); FNQ$(FND(AF))
```

540 Set step factor according to value of
absolute error

```
                        540   IF FND(AF)>180.0 THEN SF=0.1 : GOTO 565
                        545   IF FND(AF)>135.0 THEN SF=0.2 : GOTO 565
                        550   IF FND(AF)>90.0 THEN SF=0.3 : GOTO 565
                        555   IF FND(AF)>45.0 THEN SF=0.5 : GOTO 565
                        560   SF=1.0
```

565 Have we completed a trial?

`565   Pl=PA-INT(PA/NT)*NT`

570 If not, call EFIT at secondary point, and
loop

`570   IF Pl<>0 THEN PRINT : GOSUB 8845 : GOTO 475`

575 Trial completed; what next?

```
                        575   Q$="Continue, new values, stop (C/N/S)"
                        580   PRINT : PRINT Q$; : INPUT A$
                        585   IF A$="S" OR A$="s" THEN STOP
                        590   IF A$="C" OR A$="c" THEN GOTO 605
                        595   IF A$="N" OR A$="n" THEN GOTO 610
```

600 Answer not recognised; get another one

`600   PRINT "What ? " : GOTO 575`

605 Carry on with the next trial

`605   PRINT : GOSUB 8845 : GOTO 475`

610 New values required; use old values as
defaults . . .

`610   NN%=2*NO%`

615 . . . converting first to degrees

```
                        615   FOR I%=1 TO NN% : VO(I%)=FND(VO(I%)) : NEXT I%
                        620   IX=FND(IX) : AX=FND(AX)
                        625   RX=FND(RX) : OX=FND(OX) : PRINT : GOTO 70
```

```
INCLUDE DEFAULT, YESNO, JULDAY, CALDAY, ANOMALY, SUN, EFIT
```

Example

```
Fitting elliptical elements
---------------------------


How many observation to fit (3-6) [0]              ? 3

Please input the observations:

Observation number 1
Epoch of observation (D,M,Y) [0,0,0]               ? 27,4,1984
Ecliptic longitude (degrees) [0]                   ? 342.010039
Ecliptic latitude (degrees) [0]                    ? 14.860911

Observation number 2
Epoch of observation (D,M,Y) [0,0,0]               ? 27,10,1984
Ecliptic longitude (degrees) [0]                   ? 334.761506
Ecliptic latitude (degrees) [0]                    ? 2.179792

Observation number 3
Epoch of observation (D,M,Y) [0,0,0]               ? 27,4,1985
Ecliptic longitude (degrees) [0]                   ? 24.199906
Ecliptic latitude (degrees) [0]                    ? -12.604639
```

```
Your values are
----------------

Observation number 1
Epoch of observation (D,M,Y)                    27  4   1984
Ecliptic longitude (degrees)                    342.010039
Ecliptic latitude (degrees)                     14.860911
Observation number 2
Epoch of observation (D,M,Y)                    27 10   1984
Ecliptic longitude (degrees)                    334.761506
Ecliptic latitude (degrees)                     2.179792
Observation number 3
Epoch of observation (D,M,Y)                    27  4   1985
Ecliptic longitude (degrees)                    24.199906
Ecliptic latitude (degrees)                     -12.604638

Is this correct (Y or N) ? Y

Epoch for elements (D,M,Y) [0,0,0]            ? 27,10,1984

Please input starting values:

Mean anomaly at the epoch (deg.) [0]         ? 240
Mean distance (AU) [3]                        ? ,
Inclination of orbit (degrees) [30]          ? ,
Argument of perihelion (degrees) [0]         ? ,
Longitude of ascending node (deg.) [180]     ? ,
Eccentricity (0 to 1) [0.2]                  ? ,
How many passes per trial [5]                ? 6

Your values are
----------------

Epoch for elements (D,M,Y)                      27 10   1984
Mean anomaly at the epoch (deg.)                240
Mean distance (AU)                              3
Inclination of orbit (degrees)                  30
Argument of perihelion (degrees)                0
Longitude of ascending node (deg.)              180
Eccentricity (0 to 1)                           0.2
How many passes per trial                       6

Is this correct (Y or N) ? Y

*

Results after pass  0

Mean daily motion (degrees)                     0.189688
Mean anomaly at the epoch (deg.)                240
Mean distance (AU)                              3
Inclination of orbit (degrees)                  30
Argument of perihelion (degrees)                0
Longitude of ascending node (deg.)              180
Eccentricity (0 to 1)                           0.2
Step factor                                     0
Absolute error (degrees)                        86.119567

Continue, new values, stop (C/N/S)? C

*************
```

8100 PFIT and 8800 EFIT

```
Results after pass   1

Mean daily motion (degrees)                    0.164347
Mean anomaly at the epoch (deg.)               222.674481
Mean distance (AU)                             3.300957
Inclination of orbit (degrees)                 28.123064
Argument of perihelion (degrees)               332.472131
Longitude of ascending node (deg.)             181.2736
Eccentricity (0 to 1)                          1E-03
Step factor                                    0.5
Absolute error (degrees)                       42.123337

*************

Results after pass   2

Mean daily motion (degrees)                    0.223703
Mean anomaly at the epoch (deg.)               218.387578
Mean distance (AU)                             2.687614
Inclination of orbit (degrees)                 33.758421
Argument of perihelion (degrees)               328.010772
Longitude of ascending node (deg.)             177.73279
Eccentricity (0 to 1)                          0.156337
Step factor                                    1
Absolute error (degrees)                       6.222824

*************

Results after pass   3

Mean daily motion (degrees)                    0.210337
Mean anomaly at the epoch (deg.)               261.065015
Mean distance (AU)                             2.800296
Inclination of orbit (degrees)                 34.69153
Argument of perihelion (degrees)               297.75477
Longitude of ascending node (deg.)             173.16091
Eccentricity (0 to 1)                          0.205319
Step factor                                    1
Absolute error (degrees)                       3.040793

*************

Results after pass   4

Mean daily motion (degrees)                    0.214071
Mean anomaly at the epoch (deg.)               247.642337
Mean distance (AU)                             2.767636
Inclination of orbit (degrees)                 34.751695
Argument of perihelion (degrees)               311.27509
Longitude of ascending node (deg.)             173.14299
Eccentricity (0 to 1)                          0.226831
Step factor                                    1
Absolute error (degrees)                       1.19229

*************
```

```
Results after pass  5

Mean daily motion (degrees)              0.213649
Mean anomaly at the epoch (deg.)         249.041192
Mean distance (AU)                       2.771279
Inclination of orbit (degrees)           34.784293
Argument of perihelion (degrees)         309.939391
Longitude of ascending node (deg.)       173.136746
Eccentricity (0 to 1)                    0.23301
Step factor                              1
Absolute error (degrees)                 8.352E-03

*************

Results after pass  6

Mean daily motion (degrees)              0.213553
Mean anomaly at the epoch (deg.)         249.13269
Mean distance (AU)                       2.772115
Inclination of orbit (degrees)           34.788588
Argument of perihelion (degrees)         309.88912
Longitude of ascending node (deg.)       173.135711
Eccentricity (0 to 1)                    0.23326
Step factor                              1
Absolute error (degrees)                 3.6E-05

Continue, new values, stop (C/N/S)? S
```

The elements converted upon represent best values for the equinox of (average) date. Use RELEM to convert to another equinox if necessary before comparing with the original values.

LIST OF VARIABLES

The variables used in all the subroutines are listed in alphabetical order below. Each variable is shown in the left-hand column, and the names of the subroutines in which it appears are listed opposite. Variables are classified as *local*, meaning that they are used internally by the named routines and do not need to be set to any value beforehand, *global*, signifying that the variable is used again and its value must not be altered, *input* for values which need to be passed to the routine by the calling program, and *output* for the results of the calculations. A brief note about the variable is also given.

```
A        EFIT, PFIT, RELEM, DISPLAY, MOONNF, MOON, PLANS, SUN,
         ANOMALY, RISET, REFRACT, PARALLX, GENCON, EQGAL, EQECL,
         NUTAT, OBLIQ, HRANG, EQHOR, CALDAY, JULDAY, MINSEC: local
A$       CALDAY, MINSEC, YESNO: local
A0       PELMENT: local
A1       MOONNF, PELMENT, SUN: local
A1$      DEFAULT: local
A2       MOONNF, PELMENT: local
A2$      DEFAULT: local
A3       PELMENT: local
A3$      DEFAULT: local
AA       EFIT, PFIT, MOONRS, PELMENT, SUNRS: local
AB       MOONRS, SUNRS: local
AD       MOONRS, SUNRS, RISET: output:
            azimuth of setting (radians)
AE       SUN: global: Sun's eccentric anomaly
         ANOMALY: output: eccentric anomaly (radians)
AF       EFIT, PFIT: output: absolute error (radians)
AI       EFIT: local: mean anomaly at the epoch (radians)
AM       EFIT, ELOSC, PLANS: global:
            mean anomaly (input to ANOMALY; radians)
         SUN: global: Sun's mean anomaly
         ANOMALY: input: mean anomaly (radians)
AN       EQGAL: local reserved:
            gal. long. node of gal. plane on equator (rad)
AP(8)    PLANS: local: planetary anomalies
AR       EFIT, PFIT, PCOMET: input:
            argument of perihelion (radians)
AT       EFIT, ELOSC, PLANS: global:
            true anomaly (output from ANOMALY; rad)
         SUN: global: Sun's true anomaly
         ANOMALY: output: true anomaly (radians)
```

240

AU	MOONRS, SUNRS, RISET: output: azimuth of rising (radians)
AX	ELOSC: input: mean anomaly at the epoch (degrees) EFIT: input/output: estimate of mean anomaly at epoch (radians)
AZ	DISPLAY: local: azimuth of event
B	EFIT, PFIT, RELEM, DISPLAY, MOONNF, MOON, PELMENT, SUN, RISET, REFRACT, PARALLX, EQGAL, NUTAT, EQHOR, CALDAY, JULDAY: local
B$	CALDAY, MINSEC: local
B1	MOONNF, SUN: local
BM	ECLIPSE, MOONRS: global: output from MOON MOON: output: Moon's geocentric ecliptic latitude (radians)
BP	EFIT, PFIT: local: geocentric ecliptic latitude (radians) PCOMET, ELOSC, PLANS: output: geocentric ecliptic latitude (radians)
BY	ECLIPSE: local
BZ	ECLIPSE: local
C	EFIT, PFIT, MOONNF, MOON, PELMENT, RISET, EQGAL, CALDAY, JULDAY: local
C$	DISPLAY, CALDAY, MINSEC: local string
C1	PFIT, PCOMET, SUN, PARALLX, PRCESS2: local
C2	PFIT, PCOMET, PARALLX, PRCESS2: local
C3	PFIT, PCOMET, PRCESS2: local
CA	PLANS, RISET, EQGAL: local
CC(3)	GENCON: input: conversion types
CD	RISET: local
CE	GENCON, EQECL: local reserved: cos(obliquity)
CF	RISET, GENCON, EQHOR: local reserved: cos(geog. latitude)
CH	RISET, EFIT, RELEM, ELOSC, PLANS: local
CJ	RELEM: local
CL	EQGAL: local reserved: cos(gal. pole declination)
CN	ECLIPSE, PRCESS2, GENCON: local
CO	EFIT, RELEM, ELOSC, PLANS: local
CP	PARALLX, EQHOR: local
CQ	EQHOR: local
CS	GENCON: local reserved: cos(local sidereal time)
CV(3)	PRCESS2, GENCON: local: column vector
CX	PARALLX, EQECL, EQHOR: local
CY	RISET, PARALLX, EQGAL, EQECL, EQHOR: local
D	PFIT, PCOMET, ANOMALY, RISET, EQGAL, TIME, CALDAY, JULDAY: local
D$	MINSEC: local
D0	PCOMET, RELEM, ELOSC, MOONNF: local ECLIPSE: output: day number of date of eclipse
D1	SUN, NUTAT: local
D1$	DEFAULT: local
D2	SUN, NUTAT: local
D2$	DEFAULT: local
D3	SUN: local
D3$	DEFAULT: local
DA	PRCESS2, PRCESS1: input: day number of calendar date of epoch 1
DB	PRCESS2, PRCESS1: input: day number of calendar date of epoch 2
DC	RELEM: local
DD	ELOSC, ECLIPSE, MOONNF: local
DE(12,6)	EFIT, PFIT: local: derivatives matrix
DF	ECLIPSE: local
DH	ELOSC: input: day number of epoch
DI	PCOMET: input: day number of epoch of perihelion MOONRS, SUNRS: global: input to RISET RISET: input: vertical displacement at horizon (radians)

List of variables

DJ	EFIT, PFIT, PCOMET, RELEM, ELOSC, ECLIPSE, MOONNF, MOON, PLANS, PELMENT, SUN, PRCESS2, PRCESS1, NUTAT, OBLIQ, TIME: global:
	    Julian days since 1900 January 0.5
	   CALDAY: input: Julian days since 1900 Jan 0.5
	   JULDAY: output: Julian days since 1900 Jan 0.5

DK	PRCESS1: local: Julian days since 1900 Jan 0.5 of epoch 1
DM	ECLIPSE: local
DM(6)	EFIT, PFIT: local array
DN	SUNRS: local
DO	NUTAT: output: nutation in obliquity (decimal degrees) OBLIQ: global: nutation in obliquity
DP	ECLIPSE, MOONRS, SUNRS: global: output from NUTAT NUTAT: output: nutation in longitude (decimal degrees)
DQ	PCOMET: local

DS	SUNRS, TIME: input:
	   daylight saving (hours ahead of zone time)
	 DISPLAY: local:
	   daylight saving (hours ahead of zone time)

DT$	CALDAY: output: date string
DU	RELEM: input: day number of date at which elements are specified
DV	EFIT, PFIT: local
DW	EFIT, PFIT: local
DX	EFIT, PFIT, PARALLX: local
DY	PRCESS2, PRCESS1: local: argument of JULDAY PCOMET, RELEM, ELOSC, ECLIPSE, MOONNF, SUN, TIME, JULDAY: input: day number of date CALDAY: output: day number of date
DZ	ELOSC: local
E	DISPLAY, MOONNF, MOON, GENCON, EQECL, YESNO: output: 1=yes, 0=no
E1	PCOMET, RELEM, MOONNF, SUN: local
E2	MOON: local
EC	EFIT, PLANS: global: eccentricity (input to ANOMALY) ELOSC, ANOMALY: input: eccentricity SUN: global: Earth's orbital eccentricity
EH(2,6)	EFIT, PFIT: local: Earth data (ecliptic longitude, radius vector)
EM	EFIT: input: maximum allowed eccentricity
EO(12)	EFIT, PFIT: local: geocentric ecliptic coordinates based on current estimates

EP	EFIT, PFIT: local:
	   geocentric ecliptic longitude (radians)
	 PCOMET, ELOSC, PLANS: output:
	   geocentric ecliptic longitude (radians)

ER(1)	JULDAY: error flag: 0=OK, 1=impossible date
ER(2)	TIME: error flag: 0=OK, 1=ambiguous conversion
ER(3)	RISET: error flag: 0=OK, 1=no rise, -1=no set
ER(4)	SUNRS: error flag: 0=OK, 1=ambiguous sunrise/start twilight
ER(5)	SUNRS: error flag: 0=OK, 1=ambiguous sunset/end twilight
ER(6)	MOONRS: error flag: 0=OK, 1=moonrise occurs next day
ER(7)	MOONRS: error flag: 0=OK, 1=moonset occurs next day
ER(8)	ECLIPSE: error flag: 0=OK, 1=no eclipse, 2=eclipse seen elsewhere
ER(9)	ECLIPSE: error flag: 0=OK, 1=no umbral phase, 2=no total phase (lunar)
ER(10)	PFIT: error flag: 0=OK, 1=singular matrix
ER(11)	EFIT: error flag: 0=OK, 1=singular matrix
ET	RELEM: local
ET$	DISPLAY, ECLIPSE: input: type of eclipse ("L" or "S")
EX	EFIT: input/output: estimate of the eccentricity
EZ	ELOSC: local

```
F        MOONNF: local
FB       ECLIPSE: global: output from MOONNF
         MOONNF: output:
            argument of full Moon's latitude (radians)
FD       CALDAY: output: fractional part of DY
FF       ECLIPSE: global: output from MOONNF
         MOONNF: output: fractional part of JDN for full Moon
FI       ECLIPSE: global: output from MOONNF
         MOONNF: output: integer part of JDN for full Moon
FL(1)    JULDAY: flag:
            set on each call to routine; indicates same date
FL(2)    CALDAY: flag:
            set on each call to routine; indicates same date
FL(3)    TIME: flag:
            set on each call to routine; indicates same date
FL(4)    RISET, EQHOR, GENCON: flag:
            set on each call to routine; indicates same place
FL(5)    OBLIQ: flag:
            set on each call to routine; indicates same date
FL(6)    NUTAT: flag:
            set on each call to routine; indicates same date
FL(7)    EQECL, GENCON: flag:
            set on each call to routine; indicates same date
FL(8)    EQGAL: flag: set on first call to routine
FL(9)    GENCON: flag:
            set on first call; indicates same local sidereal time
FL(10)   PRCESS1: flag: set on first call; indicates same epochs
FL(11)   PRCESS2: flag: set on first call; indicates same epochs
FL(12)   PARALLX: flag:
            set on each call to routine; indicates same place
FX       DISPLAY: local

G        MOON, CALDAY: local
G1       MOONRS: local
G2       MOONRS: local
GD       MOONRS, SUNRS, EQGAL: local
GH       TIME: output: GST hours part
GL       DISPLAY, HRANG, TIME: input:
            geographical longitude (decimal degrees)
GM       TIME: output: GST minutes part
GP       DISPLAY, RISET, PARALLX, GENCON, EQHOR: input:
            geographical latitude (radians)
GR       EQGAL: local reserved:
            RA of north galactic pole (radians)
GS       TIME: output: GST seconds part
GU       MOONRS, SUNRS: local

H        DISPLAY, RISET: local
H0       ECLIPSE: local
H1       SUN: local
HD       ECLIPSE: local
HL(3)    PRCESS2, GENCON: local: work vector
HP       ECLIPSE: global: input to PARALLX
         PARALLX: input: horizontal parallax (radians)
HT       DISPLAY, PARALLX: input:
            height above sea-level (Earth radii)
HY       ECLIPSE: local
HZ       ECLIPSE: local

I        DISPLAY, PELMENT, PRCESS2, GENCON, CALDAY: local
I%       EFIT, PFIT: local
I1       PCOMET, ELOSC: local
I1%      EFIT, PFIT: local
ID       CALDAY: output: integer part of DY
II       GENCON: output: (pointer to last matrix) + 1
```

List of variables

II%	EFIT, PFIT: local	
IJ%	EFIT, PFIT: local	
IN	EFIT, PFIT, PLANS: local: inclination (radians)	
	PCOMET, ELOSC: input: inclination (radians)	
IP	PLANS: global: planet number (output from PELMENT)	
	PELMENT: output: planet number (1-8, 0=error)	
IX	EFIT, PFIT: input/output:	
	estimate of the inclination (radians)	
J	PLANS, PELMENT, PRCESS2, GENCON: local	
J0	MOONNF: local	
J1	PLANS: local: argument in perturbation terms	
J2	PLANS: local: argument in perturbation terms	
J3	PLANS: local: argument in perturbation terms	
J4	PLANS: local: argument in perturbation terms	
J5	PLANS: local: argument in perturbation terms	
J6	PLANS: local: argument in perturbation terms	
J7	PLANS: local: argument in perturbation terms	
J8	PLANS: local: argument in perturbation terms	
J9	PLANS: local: argument in perturbation terms	
JA	PLANS: local: argument in perturbation terms	
JB	PLANS: local: argument in perturbation terms	
JC	PLANS: local: argument in perturbation terms	
JI%	EFIT, PFIT: local	
JJ%	EFIT, PFIT: local	
JK%	EFIT, PFIT: local	
JL%	EFIT, PFIT: local	
K	PCOMET, ELOSC, MOONNF, MOONRS, PLANS, SUNRS: local	
KI	PFIT: local: epoch of perihelion (Julian date)	
KP(4,6)	EFIT, PFIT: input:	
	epochs of the observations (DY,MN,YR,DJ)	
KX	PFIT: input/output: estimate of epoch of perihelion (JD)	
L	PFIT, PCOMET, MOON, SUN: local	
L0	PCOMET, ELOSC, PLANS: output:	
	heliocentric ecliptic longitude (radians)	
L1	EFIT, ELOSC, PLANS, NUTAT: local	
L2	EFIT, ELOSC, PLANS, NUTAT: local	
LA	MOONRS, SUNRS: local	
LB	MOONRS, SUNRS: local	
LC	PFIT, PCOMET: local	
LD	MOONRS, SUNRS: global: output from RISET	
	RISET: output: local sidereal time of setting (hours)	
LE$	ECLIPSE: local string ("Lunar ")	
LG	EFIT, PFIT: local: Earth's ecliptic longitude (radians)	
	PCOMET, ELOSC, PLANS: output:	
	Earth's ecliptic longitude (radians)	
LH	TIME: output: local time hours part	
LI	PCOMET, ELOSC, PLANS: local: light travel time (days)	
LJ	ECLIPSE	
LL	EFIT, PFIT, PCOMET, ELOSC, PLANS: local	
LM	TIME: output: local time minutes part	
LO	EFIT, ELOSC, PLANS: local	
LP	ELOSC, PLANS: local	
LS	HRANG, TIME: output: local time seconds part	
LU	MOONRS, SUNRS: global: output from RISET	
	RISET: output: local sidereal time of rising (hours)	
LW(6)	EFIT, PFIT: local matrix	
LY	ECLIPSE: local	
M	ANOMALY: local	
M0	PCOMET, RELEM, ELOSC, MOONNF: local	
M1	MOON, SUN, PRCESS1, NUTAT, JULDAY: local	
M2	MOON, PRCESS1, NUTAT: local	
M3	MOON: local	
M4	MOON: local	

244

M5	MOON: local
M6	MOON: local
MA	PRCESS2, PRCESS1: input: month number of calendar date of epoch 1
MB	PRCESS2, PRCESS1: input: month number of calendar date of epoch 2
MD	MOON: output: Moon's mean anomaly (radians)
ME	MOON: output: Moon's mean elongation (radians)
MF	MOON: output: Moon's mean distance from its asc. node (radians)
MG	ECLIPSE: output: magnitude of eclipse
MH	ELOSC: input: epoch month number
MI	PCOMET: input: month number of epoch of perihelion
ML	MOON: output: Moon's mean longitude (radians)
MM	ECLIPSE, MOONRS: global: output from MOON MOON: output: Moon's geocentric ecliptic longitude (radians)
MN	PCOMET, RELEM, ELOSC, MOONNF, SUN, TIME, JULDAY: input: month number of date PRCESS2, PRCESS1: local: argument of JULDAY CALDAY: output: month number of date
MP	PRCESS1: local reserved: precessional constant
MR	ECLIPSE: local
MS	ELOSC, MOON, PLANS: output: Sun's mean anomaly (radians)
MT(3,3)	PRCESS2: local: precession matrix GENCON: local: coordinate conversion matrix
MU	RELEM: input: month number of date at which elements are specified
MV(3,3)	PRCESS2: local: precession matrix
MW(6,6)	EFIT, PFIT: local
MY	ECLIPSE: local
MZ	ECLIPSE: local
N	DEFAULT: input: number of parameters to get (N=1,2,3 or 0 for string)
N1	PRCESS1, NUTAT: local
N2	PRCESS1, NUTAT: local
NA	MOON: output: longitude of Moon's ascending node (radians)
NB	ECLIPSE: global: output from MOONNF MOONNF: output: argument of new Moon's latitude (radians)
NC	MINSEC: input: no. of characters in OP$ if S1=-1
ND	EFIT: local: mean daily motion (radians) ELOSC: input: mean daily motion (degrees)
NF	ECLIPSE: global: output from MOONNF MOONNF: output: fractional part of JDN for new Moon
NI	ECLIPSE: global: output from MOONNF MOONNF: output: integer part of JDN for new Moon
NM$	PELMENT: local
NN%	EFIT, PFIT: local
NO%	EFIT, PFIT: input: number of observations to fit
NP	PRCESS1: local reserved: precessional constant
NU	PFIT, PCOMET: local
NY	PRCESS1: local: number of years between epochs 1 and 2
O1	RELEM: local
OB	GENCON, EQECL: global: obliquity of the ecliptic OBLIQ: output: value of obliquity of ecliptic (decimal degrees)
OM	EFIT, PFIT, PLANS: local: longitude of the ascending node (radians) PCOMET, ELOSC: input: longitude of the ascending node (degrees)
OP$	DISPLAY, TIME: global: output from other routines MINSEC: output: answer string
OT	RELEM: local

List of variables

```
OX        EFIT, PFIT: input/output:
             estimate of the long. of asc. node (radians)
OX%       DISPLAY: local: X coordinate of centre of screen
OY%       DISPLAY: local: Y coordinate of centre of screen

P         RELEM: output: reduced inclination (radians)
          DISPLAY, ECLIPSE, MOONRS, SUNRS: global:
             output from other routines
          PARALLX: output: apparent or true hour angle (radians)
          PRCESS2, PRCESS1: output: right ascension (radians)
          GENCON: output: RA, HA or longitude (radians)
          EQGAL, EQECL: output:
             ecliptic longitude or right ascension (radians)
          HRANG: output: hour angle or right ascension (hours)
          EQHOR: output: hour angle or azimuth (radians)
P$        PELMENT: input: name of planet
P0        PCOMET, ELOSC, PLANS: output:
             radius vector of planet (AU)
P1        DISPLAY, PARALLX: local
P2        PARALLX: local
P3        ECLIPSE: local
PA        EFIT, PFIT: input/output: pass number
PD        EFIT, ELOSC, PLANS: local
PE        ELOSC: input: longitude of the perihelion (degrees)
PI        EFIT, PFIT, PCOMET, RELEM, ELOSC, DISPLAY, ECLIPSE,
          PLANS, PRCESS2, GENCON, EQGAL, EQECL: global reserved:
             value of pi
PJ        ECLIPSE: local
PL(8,9)   PLANS: global:
             planetary orbital elements (output from PELMENT)
          PELMENT: output: planetary orbital elements
PM        ECLIPSE, MOONRS: global: output from MOON
          MOON: output: Moon's horizontal parallax (radians)
PR        REFRACT: input: atmospheric pressure (mbar)
PS        PFIT, PCOMET, ELOSC, ECLIPSE, PLANS: local
PV        EFIT: local
          ELOSC, PLANS: output:
             radius vector (corrected for light time)
PX        ELOSC: input: semi-major axis (or mean distance; AU)

Q         DISPLAY: global: output from other routines
          RELEM: output: reduced argument of perihelion (radians)
          ECLIPSE, MOONRS, SUNRS: global:
             output from other routines
          MOON: local
          REFRACT: output: apparent or true altitude (radians)
          PARALLX: output: apparent or true declination
          PRCESS2, PRCESS1: output: declination (radians)
          GENCON: output: declination or latitude (radians)
          EQGAL: output: galactic latitude or declination (radians)
          EQECL: output: ecliptic latitude or declination (radians)
          EQHOR: output: declination or altitude (radians)
Q$        DEFAULT: input: prompt string
          DISPLAY: global: input prompt string to other routines
          ECLIPSE, TIME: local
          YESNO: input: prompt string
Q1        DISPLAY, PARALLX: local
Q2        PARALLX: local
QA        PLANS: local: perturbation in longitude
QB        PLANS: local: perturbation in radius vector
QC        PLANS: local: perturbation in mean longitude
QD        PLANS: local: perturbation in eccentricity
QE        PLANS: local: perturbation in mean anomaly
QF        PLANS: local: perturbation in semi-major axix
QG        PLANS: local:
             perturbation in heliocentric ecliptic latitude
```

QM	EFIT: input: maximum allowed value of mean distance (AU)
QP	EFIT, PFIT: local:' perihelion distance (AU)
	PCOMET: input: perihelion distance (AU)
QQ	PFIT: local
QW	PFIT: local
QX	EFIT: input/output: estimate of the mean distance (AU)
	PFIT: input/output:
	estimate of the perihelion distance (AU)
R	PFIT, PCOMET, ECLIPSE: local
	RELEM: output:
	reduced longitude of ascending node (radians)
R%	DISPLAY: local: radius in screen units
R0	TIME: local
R1	REFRACT, TIME: local
R1%	DISPLAY: local
R2	REFRACT: local
RC	PARALLX: local reserved:
	related to the figure of the Earth
RD	EFIT, PFIT, PCOMET, ELOSC, PLANS: local
RE	EFIT, PFIT, PCOMET, ELOSC, PLANS: local:
	radius vector of the Earth
RF	REFRACT: output: atmospheric refraction (radians)
RH	PCOMET, ELOSC, PLANS: output:
	distance from Earth, corrected for light time
RM	DISPLAY: input: radius of Moon
	ECLIPSE: output:
	radius of Moon (in equivalent hours of time)
RN	DISPLAY: input: radius of Sun
	ECLIPSE: output:
	radius of Sun (in equivalent hours of time)
RO	DISPLAY: local: ecliptic rotation angle (degrees)
RP	DISPLAY: input: radius of penumbra
	ECLIPSE: output:
	radius of penumbra (in equivalent hours of time)
	PARALLX: local
RR	EFIT, PFIT: local
	PCOMET, ELOSC, ECLIPSE, PLANS: global:
	Sun - Earth distance (output from SUN; AU)
	SUN: output: Sun - Earth distance (AU)
RS	PARALLX: local reserved:
	related to the figure of the Earth
RU	DISPLAY: input: radius of penumbra
	ECLIPSE: output:
	radius of penumbra (in equivalent hours of time)
RW	DISPLAY: local: rotation switch
RX	EFIT, PFIT: input/output:
	estimate of the arg. of perihelion (radians)
S	PFIT, PCOMET, EQGAL, EQECL: local
S$	MINSEC: output: sign string variable
S0	PCOMET, ELOSC, PLANS: output:
	heliocentric ecliptic latitude (radians)
S1	PFIT, PCOMET, MOON, PARALLX, PRCESS2: local
S2	PFIT, PCOMET, ECLIPSE, MOON, PARALLX, PRCESS2: local
S3	PFIT, PCOMET, MOON, PRCESS2: local
S4	MOON: local
SA	PLANS, EQGAL: local
SB	ECLIPSE: local
SC	DISPLAY: local: scale factor for eclipse graphics
SD	RISET: local
SE	GENCON, EQECL: local reserved: sin(obliquity)
SE$	ECLIPSE: local string ("Solar ")
SF	EFIT, PFIT: input: step factor (0 - 1)
	RISET, GENCON, EQHOR: local reserved: sin(geog. latitude)

List of variables

```
SG        HRANG: input: GST (decimal hours)
          TIME: output: GST (decimal hours)
SG$       TIME: output: GST string variable
SH        DISPLAY: input:
            Moon's ecliptic latitude at conjunction in longitude
          ECLIPSE: output:
            Moon's ecl. lat. at conjunction (equivalent hours)
SI        RELEM: local
SJ        RELEM: local
SL        EQGAL: local reserved: sin( gal. pole declination)
SM        PRCESS2, GENCON: local
SN        MINSEC: output: sign variable = -1,0,+1
SO        EFIT, RELEM, ELOSC, PLANS: local
SP        EFIT, ELOSC, PLANS: local
SR        EFIT, PFIT: local:
            Sun's true geocentric longitude (radians)
          PCOMET, ELOSC, DISPLAY, ECLIPSE, PLANS, SUNRS: global:
            output from SUN
          SUN: output: Sun's true geocentric longitude (radians)
SS        GENCON: local reserved: sin(local sidereal time)
ST        DISPLAY: global: local sidereal time (radians)
          GENCON: input: local sidereal time (radians)
SW(1)     MINSEC: input: direction switch = -1,+1
SW(2)     TIME: input: direction switch = -1,+1
SW(3)     EQECL: input: direction switch = -1,+1
SW(4)     EQGAL: input: direction switch = -1,+1
SW(5)     PARALLX: input: direction switch = -1,+1
SW(6)     REFRACT: input: direction switch = -1,+1
SW(7)     SUNRS: input: calculation type switch = 0,1,2,3
SX        EQECL, EQHOR: local
SY        RISET, PARALLX, EQGAL, EQECL, EQHOR: local

T         MOONNF, MOON, PLANS, PELMENT, SUN, PRCESS2, NUTAT, OBLIQ,
            TIME: global: Julian centuries since 1900 Jan 0.5
T0        RELEM, TIME: local reserved: GST at 0h local civil time
T1        RELEM, PRCESS1, RELEM, MOONNF, MOON, SUN, PRCESS1,
            NUTAT: local
T3        RELEM: local
T4        RELEM: local
TA        RELEM, PRCESS2: local
TB        EFIT, PFIT, PCOMET: local
TC        ECLIPSE: local
TD$       MOONRS: output: local civil time moonset
          SUNRS: output: local civil time sunset/end twilight
TF        MOONNF: local
TH        DISPLAY: local: rotation angle of North direction
          RELEM, MOONRS: local
TL        DISPLAY: global: local sidereal time (decimal hours)
          TIME: output: local time (decimal hours)
TL$       MOONRS, SUNRS: global: output from TIME
          TIME: output: local time string variable
TM        DISPLAY, ECLIPSE, MOONRS, SUNRS: global: input to TIME
          TIME: input: local time (decimal hours)
TN        MOONNF: local
TP        EFIT, PFIT, PCOMET, RELEM, ELOSC, ECLIPSE, MOON, PLANS,
          SUN, ANOMALY, RISET, PARALLX, PRCESS2, PRCESS1, GENCON,
          EQGAL, EQECL, EQHOR: global reserved: two pi
TR        REFRACT: input: atmospheric temperature (centigrade)
TT        RELEM: local
TU$       MOONRS, SUNRS: output:
            local civil time sunrise/start twilight
TW(6)     EFIT, PFIT: local array
TY        EQECL: local
TZ        DISPLAY: global: time zone (hours)
          SUNRS, TIME: input: time zone (hours)
```

```
U        EFIT, PFIT, PARALLX: local
U1       PLANS: local
U2       PLANS: local
U3       PLANS: local
U4       PLANS: local
U5       PLANS: local
U6       PLANS: local
U7       PLANS: local
U8       PLANS: local
U9       PLANS: local
UA       PLANS: local
UB       PLANS: local
UC       PLANS: local
UD       PLANS: local
UD$      MOONRS: output: UT of moonset
         SUNRS: output: UT of sunset/end twilight
UE       PLANS: local
UE$      DISPLAY: input: UT of event (string H M S)
UF       PLANS: local
UG       PLANS: local
UH       TIME: output: UT hours part
UI       PLANS: local
UJ       PLANS: local
UK       PLANS: local
UL       PLANS: local
UM       TIME: output: UT minutes part
UN       PLANS: local
UO       PLANS: local
UP       PLANS: local
UQ       PLANS: local
UR       PLANS: local
US       TIME: output: UT seconds part
UT       EFIT, PFIT: input: UT (hours; set to zero)
         DISPLAY, ECLIPSE, MOONRS, SUNRS: global:
            input/output for other routines
         MOON, SUN: input: UT (decimal hours)
         TIME: output: UT (decimal hours)
UT$      MOONRS, SUNRS: global: output from TIME
         TIME: output: UT string variable
UU       PLANS: local
UU$      MOONRS: output: UT moonrise
         SUNRS: output: UT sunrise/start twilight
UV       PLANS: local
UW       PLANS: local
UX       PLANS: local
UY       PLANS: local
UZ       PLANS: local

V0       PCOMET, ELOSC, PLANS: output:
            distance from Earth (not corrected for light time; AU)
VA       PLANS: local
VB       PLANS: local
VC       PLANS: local
VC(12)   EFIT, PFIT: local: differences matrix
VD       PLANS: local
VE       PLANS: local
VF       PLANS: local
VG       PLANS: local
VH       PLANS: local
VI       PLANS: local
VJ       PLANS: local
VK       PLANS: local
VO(12)   EFIT, PFIT: input:
            observed geocentric ecliptic coordinates (radians)
```

List of variables

```
W          EFIT, PFIT, PCOMET, RELEM, EFIT: local
W1         PFIT, MOON: local
W2         EFIT, PFIT, MOON: local
WA(12)     EFIT, PFIT: local array
WI         EFIT, PFIT: local
WJ         EFIT, PFIT: local
WS(6)      EFIT, PFIT: local array

X          EFIT, PFIT, TIME: local
           RELEM: input: inclination (radians)
           DEFAULT: input: default value/output: returned value
           DISPLAY: local & global: input to other routines
           ECLIPSE, MOONRS, SUNRS: global: input to other routines
           RISET, PRCESS2, PRCESS1: input: right ascension (radians)
           PARALLX: input: true or apparent hour angle (radians)
           GENCON: input: RA, HA, or longitude (radians)
           EQGAL: input:
             right ascension or galactic longitude (radians)
           EQECL: input:
             right ascension or ecliptic longitude (radians)
           HRANG: input: right ascension or hour angle (hours)
           EQHOR: input: azimuth or hour angle (radians)
           MINSEC: input or output (decimal degrees/hours)
X$         DEFAULT: input: default string/output: returned string
X%         DISPLAY: local
X0         DISPLAY: input:
             UT at which Moon's ecliptic latitude is zero (hours)
           ECLIPSE: output:
             UT at which Moon's ecliptic latitude is zero (hours)
X1         ECLIPSE, PARALLX, PRCESS1: local
X2%        DISPLAY: local
X3         DISPLAY: local
X3%        DISPLAY: local
X4         DISPLAY: local
X4%        DISPLAY: local
XA         PRCESS2: local
XD         TIME: local
           MINSEC: input or output (degrees/hours)
XE%        DISPLAY: local: displacement of time scale
XH         ECLIPSE: local
XI         ECLIPSE: local
XM         TIME: local
           MINSEC: input or output (minutes)
XP         MINSEC: local
XS         TIME: local
           MINSEC: input or output (seconds)

Y          EFIT, PFIT, PCOMET, ELOSC, PLANS: local
           RELEM: input: argument of perihelion (radians)
           DEFAULT: input: default value/output: returned value
           DISPLAY, ECLIPSE, MOONRS, SUNRS: global:
             input to other routines
           RISET: input: declination (radians)
           REFRACT: input: true or apparent altitude (radians)
           PARALLX: input: true or apparent declination (radians)
           PRCESS2, PRCESS1: input: declination (radians)
           GENCON: input: declination or latitude (radians)
           EQGAL: input: declination or galactic latitude (radians)
           EQECL: input: declination or ecliptic latitude (radians)
           EQHOR: input: altitude or declination (radians)
Y%         DISPLAY: local
Y0         PCOMET, RELEM, ELOSC, MOONNF: local
Y1         REFRACT, PARALLX, PRCESS1, JULDAY: local
Y2         REFRACT: local
Y3         DISPLAY: local
Y3%        DISPLAY: local
```

```
Y4        DISPLAY: local
Y4%       DISPLAY: local
YA        PRCESS2, PRCESS1: input:
            year number of calendar date of epoch 1
YB        PRCESS2, PRCESS1: input:
            year number of calendar date of epoch 2
YD        REFRACT: local
YH        ELOSC: input: year number of epoch
YI        PCOMET: input: year number of epoch of perihelion
YR        PCOMET, PRCESS2, PRCESS1: local: argument of JULDAY
          RELEM: input:
            year number of date to which elements are reduced
          ELOSC, MOONNF, SUN, TIME, JULDAY: input:
            year number of date
          CALDAY: output: year number of date
YU        RELEM: input:
            year number of date at which elements are specified

Z         RELEM: input: longitude of ascending node (radians)
          DEFAULT: input: default value/output: returned value
Z1        DISPLAY: input: UT of maximum eclipse (hours)
          ECLIPSE: output: UT of maximum eclipse (hours)
Z2        ECLIPSE: local
Z6        DISPLAY: input: UT of first contact (hours)
          ECLIPSE: output: UT of first contact (hours)
Z7        DISPLAY: input: UT of last contact (hours)
          ECLIPSE: output: UT of last contact (hours)
Z8        ECLIPSE: output: UT of start of umbral phase (hours)
Z9        ECLIPSE: output: UT of end of umbral phase (hours)
ZA        PRCESS2: local
ZB        ECLIPSE: output: UT of end of lunar total phase (hours)
ZC        ECLIPSE: output: UT of start of lunar total phase (hours)
ZD        ECLIPSE: local
ZH        DISPLAY: input:
            distance along the ecliptic (equivalent hours)
          ECLIPSE: global:
            distance along the ecliptic (equivalent hours)
ZJ        DISPLAY: input: event to display (=Z1,Z6,Z7,Z8,Z9,ZB,ZC)
```

BIBLIOGRAPHY

The Astronomical Almanac
  Published annually jointly by US Government Printing Office, Washington, and Her Majesty's Stationery Office, London.

The Explanatory Supplement to the Astronomical Ephemeris and Nautical Almanac (1961)
  Published by Her Majesty's Stationery Office, London.
  (New edition in preparation).

Astronomical Formulae for Calculators
  by Jean Meeus (fourth edition). Willmann-Bell Inc., Virginia, USA.

Practical Astronomy with your Calculator
  by Peter Duffett-Smith (third edition 1989). Cambridge University Press, Cambridge, UK.

Textbook on Spherical Astronomy
  by W. M. Smart (1977) (sixth edition). Cambridge University Press, Cambridge, UK.

Illustrating Basic
  by Donald Alcock (1977), Cambridge University Press, Cambridge, UK

Introduction to BASIC Astronomy with a PC
  by Lawrence. Willmann-Bell Inc., Virginia, USA.

Index

Index

Index

Index

PROGRAMS AVAILABLE ON DISK

The programs and subroutines listed in this book are all
available on disk. If your computer is able to read disks, you
can save yourself hours of tedious typing, and then more time
hunting for the typing errors you've made. I am able to offer
three inch, three and a half inch, five and a quarter inch, and
eight inch sized disks in a variety of different formats,
including IBM PC, Apple II, Apple Macintosh, IBM 3740, BBC Micro,
and some CP/M formats. Please ask if yours is not in this list. I
may be able to supply others.

These disks are offered **only** as an aid to users of the book. The
programs will be substantially as listed, but with the latest
corrections incorporated as they come to light. They should run
straight away under most versions of BASIC, although you might
need to make some alterations for your particular machine. I
can usually process your order within a few days, but please
allow up to 28 days for delivery.

ORDER FORM

To: Dr. P. J. Duffett-Smith,
      Downing College,
      Cambridge CB2 1DQ,
      United Kingdom.

Please send me _____ disks of size _____ inches for my computer
of type _____
(please fill in the computer type or format required).

Name_____(Block capitals)
Address_____
_____
_____

I enclose: /Cheque drawn on a UK bank account /UK Postal Order
/Sterling Eurocheque   /International Money Order /International
Postal Order /Non-Sterling Cheque/ for £_____ (Please make
cheques and orders payable to Dr. P. J. Duffett-Smith. Please add
£6.50 equivalent if paying by a non-sterling cheque or money
order to cover the conversion charge.)

Price: For posting within the UK: £12.00 per disk.
       For posting to addresses outside the UK: £13.00 per disk.

Printed in the United States
By Bookmasters

Printed in the United States
by Baker & Taylor Publisher Services